Air Show Performers

Air shows are high-risk activities that must be conducted with careful thought toward the general public, spectators, and flying and nonflying participants to ensure that the activity is as safe as reasonably possible. The impromptu, ad hoc, unrehearsed, or unplanned must never be attempted. This book offers a holistic overview of the state of safety, including safety cultural variables, safety risk parameters, and human performance factors, in the international air show community.

The aim of this book is to close the knowledge gap on safety management in air shows. It imparts to the aviation sector and other high-risk and high-performance industries the experience and knowledge that airshow performers have gained regarding risk assessment, psychological aspects, and mindfulness techniques used for safe and effective performances. The book highlights how resilient safety culture can change the air show community's mentality to deliver safer and more spectacular air show events and promotes the culture of excellence that the air show community is wedded to. The reader will obtain a thorough understanding of safety issues in air shows.

Air Show Performers: Safety, Risk Management, and Psychological Factors is a critical read for professionals within the international air show community including nonflying participants. Its appeal extends to practitioners in aviation, health and safety, and events management.

"[...] For sure, this book will become a reference and a source of inspiration for future generations of Display Pilots."

Jacques Bothelin, French Aerobatic Jet Team Leader, Honorary Board Member European Airshow Council

Manolis Karachalios was the Hellenic Air Force's F-16 Demo Team "*ZEUS*" Display Pilot for the 2010–2012 display seasons. Dr. Karachalios holds a Master of Business Administration (MBA) in Aviation Management from Coventry University, and a Doctor of Philosophy (PhD) in Aerospace Sciences from the University of North Dakota focusing on air show safety and development.

Daniel Kwasi Adjekum has over 25 years of experience in aviation as a former Ghana Air Force squadron commander, command pilot, and air display safety director. He was also an airline pilot and is currently an aviation safety consultant and professor of aviation. He is an Internationally recognized aviation safety subject-matter expert and an International Air Transport Association (IATA) certified Safety Management Systems (SMS) implementation and control expert.

Resilient Safety

Series Editor: Manolis Karachalios

Through a multidisciplinary lens, this series compiles and synthesizes cutting-edge research, best practices, and innovative methodologies from around the globe, encouraging the adoption of resilience-building measures across the highlighted sectors. Resilient Safety paves the way for a future where systems are not only shielded against threats but are also adaptable and robust, ensuring the well-being of societies in an ever-evolving risk landscape. Each volume within the series will embark on a detailed exploration of resilience as it pertains to its specific area of focus, from the engineering marvels that ensure the safety of aerospace, aviation, and maritime operations to the medical and public health innovations designed to withstand and respond to health crises. The series delves into the intricacies of environmental protection efforts that mitigate the impact of climate change, the cybersecurity defenses that protect critical information infrastructure, and the emergency management strategies that prepare communities for disaster response and recovery. Written for a diverse readership, from industry experts and academics to policy makers and the engaged public, the books in the Resilient Safety series bridge knowledge gaps and foster a shared understanding of resilience principles.

Air Show Performers

Safety, Risk Management and Psychological Factors
Manolis Karachalios and Daniel Kwasi Adjekum

For more information about this series, please visit: https://www.routledge.com/Resilient-Safety/book-series/EIERS

Air Show Performers

Safety, Risk Management, and Psychological Factors

Manolis Karachalios and
Daniel Kwasi Adjekum

Designed cover image: Hellenic Air Force (HAF)

First edition published 2024
by CRC Press
2385 NW Executive Center Drive, Suite 320, Boca Raton FL 33431

and by CRC Press
4 Park Square, Milton Park, Abingdon, Oxon, OX14 4RN

CRC Press is an imprint of Taylor & Francis Group, LLC

ISBN: 9781032556147 (hbk)
ISBN: 9781032557144 (pbk)
ISBN: 9781003431879 (ebk)
ISBN: 9781032626147 (eBook+)

DOI: 10.1201/9781003431879

Typeset in Times
by Deanta Global Publishing Services, Chennai, India

The Open Access version of chapter 1 was funded by Manolis Karachalios.

Contents

Authors ... xi
Foreword ... xii
Preface ... xiii

Chapter 1 The Evolution of Air Show Performers ... 1

1.1 Introduction ... 1
1.1.1 Who: Air Show Performers and the Broader Air Show Community ... 1
1.1.2 Where: The Venues of Air Show Performances ... 1
1.1.3 What and Why: The Evolution of Safety, Risk Management, and Psychological Factors ... 2
1.1.4 When: The Timeline of Air Show History and the Inception of This Book ... 2
1.2 Who ... 2
1.2.1 Air Show Performer Categories: A Historical Overview ... 3
1.2.2 Air Bosses and Flying Control Committee (FCC) ... 13
1.2.3 Air Show Regulators ... 15
1.2.4 Air Show Stakeholders Association/Collaboration/Bonding ... 16
1.2.5 Diversity and Inclusion in Air Shows ... 17
1.2.6 Trailblazing Female Airs Show Performers ... 20
1.2.7 All African American Air Show Performers ... 23
1.2.8 Global Perspectives ... 24
1.3 Where ... 26
1.3.1 Soaring Beyond Borders: The Inspiring Evolution of Air Shows Internationally ... 26
1.3.2 Today's Types of Air Show Events ... 27
1.4 What ... 29
1.4.1 Risk Management: From Daredevils to Professionals ... 29
1.4.2 Current Approach to Risk Management ... 29
1.5 Why ... 30
1.5.1 Evolution of Psychological Factors Affecting Air Show Performers ... 30
1.5.2 Air Show Performers' Mindset Over the Years ... 33
1.6 When ... 34
1.6.1 Early Air Shows: A Public Display of Novel Aircraft Capabilities and Pilot Skills ... 35
1.6.2 Modern Air Shows: From Exhilarating Entertainment to Episodic Tragedies ... 38

1.6.3 Air Show Safety: A Data-Driven Perspective of Psychological Factors That Influence Resilient Safety Culture in Air Shows 38
Note 39
List of References 40
Further Reading 43

Chapter 2 Risk Management 45

2.1 Introduction 45
2.1.1 What Is a Risk? 45
2.1.2 Risk Management among Air Show Performers: An Overview 46
2.2 Risks Identified in Literature and Accident Reports (*Most Frequently Observed Risks*) 48
2.2.1 Loss of Control In-flight (LOC-I) 49
2.2.2 Vertical Maneuvers 49
2.2.3 Energy-Depleting Maneuvers 50
2.2.4 Flight into Terrain (FIT) or Flight into Object (FIO) 50
2.2.5 Mid-Air Collision (MAC) 50
2.2.6 Human Factors (HF) 50
2.2.7 Push-Pull Effect 51
2.2.8 Distractions in the Cockpit 52
2.2.9 Mechanical Failure 52
2.2.10 Bird Strikes 52
2.3 Risk Management Planning for Air Show Performers 52
2.3.1 Risk Perception 53
2.3.2 Risk Tolerance 53
2.3.3 Psychological Risk Factors Associated with the Air Show Operations 54
2.3.4 Variations in Risk Management Based on Demographics 75
2.3.5 Practical Examples of Effective Risk Management Methods 78
2.4 Enhancing Risk Management in Air Show Performances 79
2.4.1 Airshow Operational Risk Management (AORM) Framework for Performers 80
2.4.2 Systematic Approach to Risk Management in Air Shows 85
2.4.3 The Insidious Hazard Associated with the Normalization of Deviance 89
Notes 91
List of References 91

Chapter 3 Air Show Culture 96

3.1 Introduction 96
3.2 Safety Culture 97
3.3 Resilient Safety Culture 98
3.4 Air Show Performers' Excellence Culture 100
3.4.1 Defining Excellence Culture for Air Show Performers 100
3.4.2 Key Elements of Air Show Excellence Culture 100
3.4.3 Fostering a Culture of Excellence among Air Show Performers: Strategies 101
3.5 Perspectives from Air Show Performers on Air Show Culture 105
3.5.1 Interviews Results 105
3.5.2 Effect of Air Show Flying Experience on Perception of Resilient Safety Culture 111
3.6 Case Study: Fast Jet Solo Display Pilot Safety Workshop 112
3.6.1 Background 112
3.6.2 Workshop's Exclusive Features Shaping Display Pilots' Safety Culture 113
3.6.3 Recommendations for the Air Show Community 114
Note 115
List of References 115

Chapter 4 Psychological Factors 118

4.1 Introduction 118
4.2 Hazardous Attitudes 119
4.3 Biases 120
4.3.1 The 5B Model 120
4.3.1.1 Introduction 120
4.3.2 The *5B* Domino Effect 123
4.4 Study Findings 126
4.4.1 Hazardous Attitudes Identified by the FAA as Detrimental to Flight Safety 127
4.4.2 Concealed Hazardous Attitudes 128
4.4.3 Antidotes to Concealed Hazardous Attitudes 130
4.5 Psychological Effects of Physiology on Air Show Performers 131
4.5.1 G Forces 132
4.5.2 Push-Pull Effect 133
4.5.3 Low-Level Aerobatics 134
4.6 Case Study: The 2015 Shoreham Airshow Tragedy: The Role of Psychological Factors in the Pilot's Decision-Making 136
4.6.1 Introduction 136

4.6.2 Background ... 136
4.6.3 Psychological Factors Affecting the Pilot's Decision-Making ... 136
4.6.4 5B Model Analysis ... 137
4.6.5 Lessons Learned ... 138
4.6.6 Conclusion ... 140
Notes ... 140
List of References ... 140
Further Reading ... 142

Chapter 5 Mindfulness ... 143

5.1 Introduction ... 143
5.2 Mindfulness ... 143
5.2.1 Individual Mindfulness ... 144
5.2.2 Mindfulness Training ... 144
5.2.3 The Power of Mindfulness for Air Show Performers ... 146
5.2.4 Incorporating Mindfulness into Air Show Performers' Daily Routines: Practical Applications ... 147
5.2.5 Adapting Mindfulness Techniques for High G Situations in Air Show Performances ... 149
5.2.6 Maintaining Synchronized Mindfulness Among Air Show Team Members ... 150
5.3 Study Findings ... 151
5.3.1 Visualization ... 152
5.3.2 Exogenous Factor Control ... 155
5.3.3 Preshow Preparation ... 155
5.3.4 Consistency ... 156
5.3.5 Role of a Safety Observer ... 157
5.3.6 60-minute Rule ... 157
5.4 Case Studies ... 158
5.4.1 Solo Display Pilots: "Walk the Talk" Method ... 158
5.4.2 Demo Teams: Blue Angels ... 161
Notes ... 162
List of References ... 162
Further Reading ... 165

Chapter 6 Epic Survivals ... 166

6.1 Introduction ... 166
6.2 Case Study 1. Lessons from a Near-Miss Experience: Hellenic AF F-16 Demo Team *ZEUS,* 2012 (Manolis Karachalios) ... 167

6.2.1 What, When, Where: The Facts 167
6.2.2 Why: The Psychological Factors Involved 170
6.2.3 How: Risk Management Strategies that Contributed to the Epic Survival 174
6.2.4 Lessons Relearned 176
6.3 Case Study 2. Escaping a Wounded Hornet, RCAF CF-188, 2010 177
6.3.1 Interview Analysis 177
6.3.2 What, When, Where: The Facts 179
6.3.3 Why: The Psychological Factors Involved 180
6.3.4 How: Risk Management Strategies that Contributed to the Epic Survival 181
6.4 Case Study 3. Resilience and Safety Culture: The Red Arrows' Triumph Over Tragedy, RAF Red Arrows, 2010 182
6.4.1 What, When, Where: The Facts 182
6.4.2 Why: The Psychological Factors Involved 183
6.4.3 How: Risk Management Strategies that Contributed to the Epic Survival 185
6.5 Case Study 4. An Unforeseen Encounter: Epic Survival among Air Show Performers in Vintage Warbirds, 2008 187
6.5.1 What, When, Where: The Facts 187
6.5.2 Why: The Psychological Factors Involved 188
6.5.3 How: Risk Management Strategies that Contributed to the Epic Survival 190
6.6 Case Study 5. A Near Disaster at the Paris Air Show: The 1989 MIG-29 Bird Strike Incident 191
6.6.1 What, When, Where: The Facts 191
6.6.2 Why: The Psychological Factors Involved 193
6.6.3 How: Risk Management Strategies that Contributed to the Epic Survival 193
6.7 Case Study 6. Exceptional Teamwork: Wing Walker Ditching at UK, 2021 195
6.7.1 What, When, Where: The Facts 195
6.7.2 Why: The Psychological Factors Involved 196
6.7.3 How: Risk Management Strategies that Contributed to the Epic Survival 197
Notes 198
List of References 199

Chapter 7 The Evolution of Air Shows: From Thrilling Spectacle to Safe and Sustainable Exhibition 205

7.1 Introduction 205
7.2 Safety 206

7.3 Sustainability ... 206
7.3.1 Environmentally Sustainable Practices ... 207
7.3.2 Sustainable Financial Support and Sponsorship ... 208
7.3.3 Collaboration and International Partnerships ... 208
7.4 Innovation ... 210
7.4.1 Virtual and Augmented Reality ... 210
7.4.2 3D Sound and Holographic Displays ... 210
7.4.3 Hybrid Events ... 211
7.4.4 Integration of Social Media ... 211
7.4.5 Shows with Small Unmanned Aerial Vehicles (UAVs) ... 212
7.4.6 Integration of Urban Air Mobility (UAM) Demonstrations ... 213
7.5 Diversity and Inclusivity ... 214
7.5.1 Diverse Air Shows ... 214
7.5.2 Accessible and Inclusive Events ... 215
7.6 Community-Centered Experiences ... 215
7.6.1 Intergenerational Entertainment ... 215
7.6.2 Educational and Outreach Programs ... 216
7.7 The Future Air Show Performer: Embracing Innovation, Diversity, and Sustainability ... 216
7.7.1 Innovative Performances ... 217
7.7.2 Advancements in Technology ... 218
7.7.3 New Air Show View: Shifting to a Culture of Excellence as the Core Ingredient for Safety ... 219
7.7.4 Environmental Sustainability ... 220
7.8 Biking at Air Shows: Demonstration of Commitment to Sustainability and Well-Being ... 221
Notes ... 222
List of References ... 222

Chapter 8 Concluding Remarks ... 224

Glossary ... 225

Index ... 227

Authors

Manolis Karachalios, PhD, MBA, ATP, CFI.

Dr. Manolis Karachalios was the Hellenic Air Force's F-16 Demo Team "*Zeus*" Display Pilot for the 2010 to 2012 display seasons. He joined the Hellenic Air Force Academy in 1996 and graduated with honors in 2000. Upon graduation he spent 2 years as an operational A-7H Corsair pilot and then he joined the 340 Squadron, flying the F-16 Block 52+. Since 2014, he has been a civilian instructor pilot and an aviation consultant, while in 2018 he took over as the flying display director (FDD) of the Athens Flying Week International Air Show (AFW). His passion for air shows keeps him involved as a Member of the Board of the European Airshow Council (EAC) and the International Air Show Safety Team (IASST), as well as a moderator of the annual Fast Jet Solo Display Pilots' Safety Workshop.

Dr. Karachalios holds a Master of Business Administration (MBA) in Aviation Management from Coventry University, and a Doctor of Philosophy (Ph.D.) in Aerospace Sciences from the University of North Dakota focusing on air show safety and development.

Daniel Kwasi Adjekum, PhD, ATP, CSP.

Dr. Daniel Kwasi Adjekum has over 25 years of experience in aviation as a former Ghana Air Force squadron commander, command pilot, and air display safety director. He was also an airline pilot and is currently an aviation safety consultant and professor of aviation. He is an Internationally recognized aviation safety subject-matter expert and an International Air Transport Association (IATA) certified Safety Management Systems (SMS) implementation and control expert.

He holds a Master of Science in Aviation and a Doctor of Philosophy (Ph.D.) in Aerospace Sciences from the University of North Dakota. He is a certified safety professional (CSP) accredited by the Board of Safety Professionals (BCSP) of the United States. Dr. Adjekum is currently an assistant professor at the aviation department (UND) teaching courses in crew resource management, safety management systems, and human factors.

Dr. Adjekum has published extensively in peer-reviewed academic journals and presented at academic conferences and industry workshops both in Ghana and in the US. He is regularly sought after by local, national, and international media houses to provide expert insight on aviation accident investigations and relevant aviation safety issues.

Foreword

I first heard of Dr. Manolis Karachalios when I was invited to participate in a survey he designed to contribute to a deeper understanding of the safety culture in the international air show community, and I was immediately intrigued.

I have flown airshows for more than 35 years in a variety of different airplanes, around the world, in all sorts of weather conditions. Airshows are a high-risk environment, and managing those risks is my most important job. For me, it begins months before the airshow season when considering next year's schedule, maintaining my airplane, and organizing my life around my training and practice program by staying in good health and conditioning myself to maintain a good G tolerance.

For airshow pilots, psychological factors are important too. It takes practice and discipline to learn to compartmentalize and manage unwanted distractions before we fly. We have to learn to be mindful, present and focused in an extremely dynamic environment where there are many people and many demands on our time.

Safety, risk management, and the psychology of handling stress in the airshow environment are the things I have literally been obsessed with since I first began my career. Any pilot, and especially those of us who operate in a high-risk environment, must continually analyze, assess, and, importantly, learn and improve, not only for themselves but for future generations.

Dr. K's book provides a framework for continuous improvement of safety. Not just a spectator, he is an active participant in the airshow community. Dr. K's experience as a fighter pilot flying F-16s in the Hellenic Air Force, as the Flying Director for the Athens Flying Week Airshow, and as a longtime member of the Board of Directors for the European Airshow Council makes his research and insights into how we can improve our community and its safety culture, uniquely qualified and important.

Dr. K's findings give pilots like me a fresh perspective. Listening to other voices can give us a different view of our industry. And, while it's no big surprise that his findings suggest most airshow performers embrace a resilient safety culture, how we best put that into practice and manage those risks is the key.

Best safety practices in aviation and airshow flying are a lifelong commitment to learning and improving. It's thanks to Dr. Karachalios and Dr. Adjekum, and other researchers that we have been given some wonderful tools to do that with.

Patty Wagstaff
American Aerobatic Pilot, Three-time U.S. National Aerobatic Champion, Inductee of the National Aviation Hall of Fame

Preface

Picture yourself high above a bustling crowd of thousands, spectators squinting into the sun, their faces turned upwards in awe and anticipation. You're at the helm of a high-energy air show, piloting through a choreographed dance in the sky, commanding the rapt attention of every onlooker on the ground. Amid the thrill and spectacle, behind the deafening roar of the engines and the dazzling feats of aerial acrobatics, an uncompromising principle anchors everything: safety. It's in this electrifying yet demanding environment that the book, *Air Show Performers: Safety, Risk Management, and Psychological Factors*, finds its relevance and purpose.

Driven by years of hands-on experience and meticulous research, this book unravels the intricate tapestry of safety culture in the aviation industry. It's a deep dive into the complexities and dynamics of ensuring paramount safety during air shows, a spectacle that combines extreme skill with inherent risks.

The result of an ambitious study, this work seeks to illuminate the key factors that contribute to safety culture and operational risk management within the air show community. It carefully dissects the multifaceted relationships between operational risk factors, hazardous attitudes, resilient safety culture, and industry mindfulness. The outcome is a blueprint, clear and actionable, intended for the continuous enhancement of safety in the air show realm.

As you flip through the pages, you'll find it filled with insights and revelations, all distilled from rigorous fieldwork and innovative research. The narrative underscores the urgent need to foster a culture of excellence and mindfulness in the air show industry, challenging every stakeholder – performers, organizers, and regulators – to be an agent of safety.

This book is an open invitation, a call to arms, to all those who share the noble objective of advancing safety in the aviation industry. It strives to be a comprehensive resource for those committed to this critical cause. It's a call to action, a guide, and a compass, all in one.

We'd like to conclude by expressing our heartfelt appreciation to the dedicated professionals within the air show community, whose invaluable contributions have been instrumental in bringing this enlightening work to fruition. As we embark on this transformative journey together, we wish everyone safe skies and fair winds.

Manolis Karachalios and Daniel Kwasi Adjekum

1 The Evolution of Air Show Performers

1.1 INTRODUCTION

It is a cherished honor and pleasure to introduce our readers to the fascinating world of air show performance and its development over time. Throughout this introductory chapter, we use a *5W* approach (who, where, what, why, and when) to emphasize the safety, risk management, and psychological factors related to the evolution of air show performance. As a comprehensive resource for performers, organizers, and enthusiasts alike, this book explores the challenges and opportunities associated with air show performances while shedding light on the importance of maintaining a robust safety culture within the community. Moreover, it is vital to learn from past safety events and prevent the recurrence of tragic air show disasters, such as the Farnborough tragedy in 1952, the Ramstein catastrophe in 1988, the Sknyliv disaster in 2002, the Shoreham crash in 2015, and the recent Commemorative Air Force midair collision in Dallas, Texas, in 2022.

1.1.1 Who: Air Show Performers and the Broader Air Show Community

We begin by discussing the various categories of air show performers and the wider air show community throughout history. We delve into the evolution of performances from their inception to the present day and the diversity, inclusion, and demographic aspects of this unique group of individuals. Additionally, we introduce the roles and responsibilities of key stakeholders involved in air show events, such as air bosses, regulators, and air show associations.

1.1.2 Where: The Venues of Air Show Performances

In this section, we explore the different locations where air show performances have occurred over time, including the transition from early airfields to modern venues. This will include a discussion on the types of air show venues, the unique challenges they present, and the adaptations performers have made to ensure both their safety and the safety of spectators.

DOI: 10.1201/9781003431879-1

1.1.3 What and Why: The Evolution of Safety, Risk Management, and Psychological Factors

This section further examines how safety, risk management, and psychological factors have evolved alongside air show performances. We will discuss the factors contributing to human errors among air show performers and how these factors can increase the likelihood of safety events, such as incidents and accidents. Drawing on research and case studies from the aviation industry, we will provide practical strategies and recommendations for mitigating the effects of human error in air show performances. This will include discussions on training, standard operating procedures (SOPs), and the role of a safety culture within the air show community.

1.1.4 When: The Timeline of Air Show History and the Inception of This Book

The final section of this chapter recounts the timeline of air show history, from the earliest public displays of powered flight to the modern, high-energy performances we see today. We thoroughly examine critical milestones, technological advancements, and influential performers who have shaped the course of air show history, providing context for the evolution of safety, risk management, and psychological factors in air show performances.

Additionally, we discuss the motivations and objectives for authoring this book, which originate from a desire from the authors to explore and understand the complex interplay between air show performance, safety, risk management, and psychological factors and add to the existing body of knowledge. Most of the contents in this book are data-driven findings from seminal doctoral research, and it is a heartfelt desire of the authors to share these findings and recommendations with the broader aviation community.

Focusing on the development of air shows and the related evolution of safety, risk management, and psychological aspects, this chapter provides a foundation for examining the wide world of air show performers and the larger air show community. We intend to provide a thorough introduction that will engage, educate, and inspire readers as they journey through the fascinating and demanding world of aerial displays by addressing the who, where, what, why, and when of air show history as well as the conception of this book.

By providing this well-rounded introduction to the world of air shows, we hope to inform and enthrall our readers and contribute to the body of knowledge available to academia, aviation specialists, and other professionals involved in high-performance events. Ultimately, our goal is to inspire the next generation of aviation enthusiasts to explore the thrilling and rewarding world of air shows.

1.2 WHO

The awe-inspiring feats of air show performers have captivated audiences for over a century as they soar through the skies, defying gravity and showcasing the boundless potential of human innovation. Behind the breathtaking aerial ballets and the thunderous roars of aircraft engines lies a complex and fascinating world where safety, risk management, and psychological factors play a vital role in ensuring these enthralling spectacles' success and continued advancement.

Every time most air show performers take to the skies, they demonstrate their exceptional skills and their unwavering dedication to ensuring the safety of both themselves and their audience.

The importance of writing about the safety of air show performers cannot be overstated, as it serves as a reminder of the immense responsibility that these pilots bear and the extensive measures they undertake to protect everyone involved. By exploring the intricate world of air show safety, we shine a light on the remarkable achievements of those who work tirelessly behind the scenes to make these events possible while emphasizing the ongoing need for vigilance, innovation, and collaboration in maintaining the highest safety standards.

1.2.1 Air Show Performer Categories: A Historical Overview

Air shows have been a popular form of entertainment since the early days of aviation, capturing the imagination of both aviation enthusiasts and the general public alike. From daring aerial stunts to breathtaking formation flying, air show performers have evolved and diversified over the years, with each category offering a unique perspective on the capabilities of both pilots and aircraft. An overview of the various categories of air show performers is provided, examining their historical evolution and offering insights into their unique skills and talents. This section delves into the various categories of air show performers, providing a comprehensive understanding of their unique characteristics and historical evolution. By examining both academic and air show enthusiast perspectives, we offer you a holistic view of the world of air show performers and their diverse range of talents. The section also provides a rich historical foundation of air shows and the diverse background of air show performers over time.

As stated earlier, there have been a natural evolution and diversification in performance at air shows, providing more unique and captivating experiences for all stakeholders in this area of aviation.

In contemporary times, air show performers play an essential role in showcasing the complexity, capabilities, and diversity required in aerial demonstration of aircraft and other aviation-related products. From aerobatic pilots and wing-walking performers to jet demonstration teams and glider performers, air show performers use their skill, precision, and creativity to put on thrilling displays for audiences worldwide. In the pursuit of their craft, air show performers sometimes push their skills to the very edge of safe operational envelopes. An explicit adherence to regulations, guidelines, and standard operating procedures (SOPs) set forth by aviation safety regulators is vital to ensure their safety and that of their audience while inspiring the development of new aviation technology.

As we reflect on the inspiring evolution of air show performers internationally, let us remember the indomitable spirit of the pioneers who first took to the skies and the countless individuals who have dedicated their lives to the pursuit of their dreams. May their stories of courage, determination, and innovation serve as a beacon of hope for future generations of air show performers, reminding us all of the boundless potentials that lie within each of us when we dare to reach for the skies and embrace the extraordinary.

1.2.1.1 Civilian Aerobatic Performers

Civilian aerobatic performers are pilots who are not affiliated with any military organization and perform a variety of aerial maneuvers to entertain the audience. This category includes solo performers and teams flying a wide range of aircraft types, from vintage biplanes to modern high-performance aircraft. Some well-known civilian aerobatic performers include Jacques Bothelin, France (see Figure 1.1); Matt Hall, Australia; Rob Holland, USA; Selwyn "Scully" Levin, South Africa; Jorge Malatini, Argentina; Yoshihide Muroya, Japan; and Patty Wagstaff, USA.

Historically, civilian aerobatic performers have their roots in the barnstormers of the 1920s and 1930s, who would travel from town to town, putting on daring displays of aerial stunts in mostly surplus World War I aircraft, such as the Curtiss JN-4D, also known as the "*Jenny*" (see Figure 1.2).

1.2.1.2 Military Demonstration Teams

Military demonstration teams are composed of highly skilled military pilots who showcase their nation's aviation prowess, highlight the latest technological capabilities of aerial platforms, and invariably use these demonstrations as promotional avenues for recruiting personnel into their respective military branches. These teams

FIGURE 1.1 Demonstration team leader Jacques Bothelin. Copyright: Jacques Bothelin.

FIGURE 1.2 Gladys Roy, with her opponent Ivan Unger, playing tennis, on a Curtiss "Jenny" JN-4D aircraft piloted by Frank Tomac, who kept the plane at 3,000 feet – circa 1925. Copyright: Bettman Collection / Getty Images.

often perform formation flying, showcasing high precision and coordination. Notable military demonstration teams include the Royal Australian Air Force's "Roulettes," the Chinese Air Force's "Ba Yi," the Chilean Air Force "Halcones," the Royal Canadian Air Force Snowbirds, the Indian Air Force's "Surya Kiran," the Polish Air Force's "Orlik," the Royal Air Force's Red Arrows, the Russian Air Force's "Russian Knights," the South African Air Force's "Silver Falcons," the United Arab Emirates Air Force's "Fursan Al Emarat," and the United States Navy Blue Angels.

The first military demonstration team can be traced back to the Patrouille d'Étampes, a French team formed in 1931. Following World War II, several other nations established their teams, leading to the growth and popularity of military demonstration teams worldwide.

Jet demonstrations feature high-performance aircraft, such as fighter jets and training aircraft, showcasing their speed, power, and agility. The Lockheed Martin F-16 (see Figure 1.3), the Boeing F-18, the Eurofighter Typhoon, and the Saab JAS-39 Gripen are among the most frequently demonstrated fast jets in the air shows, while the Dassault Alpha Jet, the BAE Hawk, the Aero Vodochody L-39 Albatros, and the Aermachhi MB-339 are some of the training jets flown by military demonstration teams.

Even though military transport aircraft might not be the stars of an air show in terms of acrobatic stunts, they still play significant roles in these events, showcasing their astounding size, capabilities, and engineering marvel. These aircraft, for example, Airbus A400M Atlas, Boeing C-17 Globemaster III, Alenia C-27J Spartan, the

FIGURE 1.3 F-16 Fighting Falcon of the Zeus Demo Team, Hellenic Air Force, Demonstrating at RIAT 2015. Copyright: Hellenic Air Force.

Lockheed Martin C-130 Hercules, and Ilyushin Il-76, typically perform tasks such as equipment, troop, and supply transportation. While these aircraft don't execute aerobatic maneuvers like fighter jets, they can execute impressive displays, including demonstrations of their takeoff, landing, and cargo-loading abilities. Additionally, they may participate in flyovers and formation flights with other military aircraft. Such an aircraft that steals the crowds' attention is the *Fat Albert* C-130 Hercules (see Figure 1.4). This aircraft is part of the U.S. Navy's Blue Angels flight demonstration

FIGURE 1.4 C-130 Fat Albert of the Blue Angels Demonstration Team, U.S. Navy, Demonstrating at Fargo Airshow 2015. Photo by the author (Daniel Kwasi Adjekum).

squadron and frequently opens for the Blue Angels by demonstrating its impressive short takeoff and landing capabilities.

1.2.1.3 Helicopter Performers

Helicopter performers captivate audiences with their astonishing stunts, demonstrating the impressive capabilities and agility of these aerial marvels. Some of the jaw-dropping maneuvers include gravity-defying loops, where the helicopter climbs vertically before flipping upside down and descending back to its original position, creating a full circle in the sky. They also perform exhilarating rolls, where the helicopter rotates sideways around an imaginary line running from its nose to its tail. These talented pilots showcase other thrilling moves, such as steep dives toward the ground, spinning maneuvers resembling a corkscrew, and sharp, high-speed turns that demonstrate the incredible control and stability of these versatile flying machines.

Helicopter performers must adhere to stringent safety guidelines to address the unique challenges and risks associated with their demonstrations. Unlike fixed-wing aircraft, helicopters operate with a higher degree of maneuverability and agility, which can lead to more complex and unpredictable flight patterns. Key hazards include the potential for rotor blade strikes, tail rotor issues, and loss of lift due to proximity to other helicopters or ground structures. Unlike fixed-wing aircraft, helicopters have limited energy reserves in the event of an engine failure, making autorotation (emergency landing technique) more critical. Pilots should be well-versed in autorotation procedures and practice them regularly to ensure a safe emergency landing. To mitigate these risks, helicopter performers must follow specific safety measures, such as maintaining minimum separation distances and carefully planning flight paths to avoid potential obstructions. These precautions help ensure the safety of both the performers and spectators during these thrilling aerial displays. Otto the Helicopter has been a legend of the North American air show industry, while the Royal Navy Black Cats is an example of a military helicopter display team (see Figure 1.5).

1.2.1.4 Warbird Performers

Warbird performers fly historic military aircraft, often restored to their original condition, to pay homage to their role in aviation history and honor the pilots who flew them in combat. These performances often include reenactments of historical aerial battles or demonstrations of specific aircraft capabilities. Some notable warbird performers include the Commemorative Air Force in the United States, the Fighter Collection at the Imperial War Museum Duxford in the UK, and the Flying Heritage and Combat Armor Museum located in Everett, Washington, USA.

The rise of warbird performers can indeed be attributed to the preservation and restoration efforts that began in the 1960s and 1970s. As collectors and enthusiasts sought to preserve and showcase historical military aircraft, they paved the way for the growth of warbird demonstration performances. These performances allow audiences to experience the power and elegance of vintage military aircraft, such

FIGURE 1.5 The Royal Navy Black Cats Display Team performing at RIAT, UK, 2016. Copyright: Tim Felce.

as bombers and fighter planes, while also educating them on the rich history and significance of these machines.

Warbird demonstration performers are required to comply with specific safety guidelines to ensure the safety of both the pilots and the spectators. These guidelines typically include thorough aircraft maintenance protocols, regular inspections, and adherence to safety standards set by aviation regulatory bodies.

Some well-known warbird demonstration performers include the B-25 Mitchell Bomber and the P-51 Mustang. The B-25 Mitchell Bomber is a twin-engine, medium bomber that was used extensively during World War II. It gained fame for its role in the Doolittle Raid on Tokyo, which was a key turning point in the Pacific War. The P-51 Mustang, on the other hand, is an iconic single-seat fighter aircraft that was used by the Allies during World War II and the Korean War. Its exceptional range and performance made it one of the most successful fighter planes of the era.

These warbird demonstrations not only entertain but also serve as a living tribute to the brave men and women who flew and maintained these aircraft during times of conflict. By keeping these historic aircraft flying, warbird performers help to preserve an important part of aviation history for future generations to appreciate and learn from.

1.2.1.5 Glider Performers

Glider performers utilize engineless aircraft to perform graceful and elegant aerial displays, relying solely on their skill and the forces of nature to stay aloft. These performances often include aerobatics and formation flying, demonstrating the art and science of soaring flight. Some well-known glider performers include Manfred Radius, Bob Carlton, and Luca Bertossio (see Figure 1.6).

FIGURE 1.6 Luca Bertossio glider air show performer during his display at Athens Flying Week International Air Show, 2021. Copyright: Athens Flying Week (Photo by George Markakis).

The origins of glider performers can be traced back to the early 20th century when gliding became a popular recreational activity in Europe. As technology advanced and gliders became more capable, pilots began to develop and showcase their aerobatic skills in these engineless aircraft.

Glider performers showcase the capabilities and agility of gliders, including stunts such as loops and rolls. Glider performers must comply with specific safety guidelines, including minimum separation distances between aircraft and communication procedures between pilots. The British Gliding Association (BGA) guides glider display flying (*British Gliding Association*, 2023). The Airborne Pyrotechnics display team is an example of a glider performer team.

1.2.1.6 Parachute Demonstration Teams

Parachute demonstration teams are composed of highly skilled skydivers who perform various aerial stunts while descending under their parachutes. These performances often include formation skydiving, free-flying, wingsuit flying, and canopy piloting. Some renowned parachute demonstration teams include the United States Army Golden Knights, the United States Navy Leap Frogs (see Figure 1.7), and the Red Bull Air Force.

Parachute demonstration teams emerged during the 1950s and 1960s as military units began to develop and showcase their expertise in airborne operations. Over

FIGURE 1.7 U.S. Navy Leapfrogs performing at Fargo Airshow, 2015. Photograph by the author (Daniel Kwasi Adjekum).

time, civilian skydiving teams also formed, creating a diverse range of parachute demonstration teams worldwide.

Skydiving performers include both solo and team displays, with performers jumping from aircraft and performing stunts in freefall. The British Skydiving guides display skydiving, including safety considerations and qualifications for display skydivers. The Red Bull Air Force is an example of a skydiving performer team.

1.2.1.7 Paragliding Performers

Paragliding performers use lightweight, nonmotorized aircraft to perform stunts and aerial displays, including high-speed runs and spirals. The safety guidelines for paragliding performances can vary depending on the country or organization overseeing the event, but some general rules and regulations often apply.

For the British Hang Gliding and Paragliding Association (*BHPA*, 2023), the specific safety guidelines include the following:

- Minimum altitude: Display pilots must maintain a minimum altitude above the ground, usually between 100 and 500 feet (30–150 meters), depending on the complexity of the stunts and the location. Exact minimum altitudes might be specified by the event organizer, local aviation authorities, or the organization overseeing the event.
- Pilot qualifications: Display pilots should have the appropriate license or certification level to perform in an event. This often means holding an advanced pilot rating from a recognized paragliding organization such as the BHPA, the United States Hang Gliding and Paragliding Association (USHPA), or other national associations. Additionally, display pilots are usually required to have a significant amount of experience, including a minimum number of logged flights and/or hours in the air.

The Red Bull X-Alps, as a paragliding race and display event, would also adhere to safety guidelines set forth by the event organizers and the appropriate paragliding organizations. It is essential for pilots participating in such events to be highly skilled, experienced, and have a deep understanding of safety protocols.

1.2.1.8 Wing-Walking Performers

Wing-walking performers perform stunts on the wings of aircraft in flight, including handstands, acrobatics, and even transferring between planes mid-flight. Wing-walking performers use specially designed harnesses and may have additional safety equipment, such as life vests and helmets. AeroSuperBatics is one example of a wing-walking performer team (see Figure 1.8).

1.2.1.9 Radio-Controlled (RC) Aircraft Performers

The history of RC aircraft performances dates back to the 1930s when hobbyists began building and flying their model aircraft. As technology advanced, so did the capabilities and performance of these aircraft, leading to the growth of the RC aircraft performing community.

RC aircraft performers operate scale model aircraft, demonstrating exceptional piloting skills and precise aircraft control. These performances can include aerobatics, formation flying, and even reenactments of historical aerial engagements. Some notable RC aircraft performers include Jase Dussia, Frank Tiano, and the RC Kavala Acro Team (see Figure 1.9).

1.2.1.10 Drone Performers

Drone performers have emerged as a relatively recent addition to the world of air shows. These performers operate unmanned aerial vehicles (UAVs) or drones

FIGURE 1.8 AeroSuperBatics Wingwalkers. Copyright: Tim Felce.

FIGURE 1.9 RC Kavala Acro Team is performing a vertical hovering of their RC Yak 54 model. Copyright: RC Kavala Acro Team.

equipped with LED lights, allowing for synchronized aerial light displays and choreographed performances.

Drone performances began to gain popularity in the early 2010s as advances in drone technology allowed for greater stability, precision, and control. In recent years, drone performances have garnered significant attention for their innovative and mesmerizing displays for swarms of drones, pushing the boundaries of what is possible in aerial performances.

1.2.1.11 Mix Performances

Mix performances have developed alongside the growth and diversification of air show performers, offering a fresh approach to aerial entertainment. These performances showcase the creativity and adaptability of air show performers as they continuously seek new ways to captivate and inspire their audiences.

Mix performances combine multiple categories of air show performances, creating a unique and engaging experience for the audience. Examples of mixed performances included the Red Bull Air Race, which combines aerobatics and timed racing through a challenging course of inflatable pylons, and the jet trucks racing against aircraft (see Figure 1.10).

1.2.1.12 Aircraft Innovations

Sustained interest and continuous improvements in air show performance have inspired cutting-edge designs and innovations in aircraft designs. An example is the

FIGURE 1.10 Jet truck racing against an aerobatic aircraft at Cleveland National Air Show, USA. Copyright: Erik Drost.

Jetman wing, developed by Swiss inventor Yves Rossy, which allows a pilot to fly at speeds of up to 250 mph using a jet-powered wing. Such innovations contribute to the development of aviation technology and add to the excitement and spectacle of air shows.

1.2.2 Air Bosses and Flying Control Committee (FCC)

Air bosses and flying display directors play a vital role in ensuring the safety and success of air shows. They are responsible for managing the complex logistics of air shows, including air traffic management, designing and coordinating aerial displays, and managing risks during the event (see Figure 1.11). Air bosses and flying display directors must have extensive knowledge and experience in aviation and air traffic management to ensure the display is performed safely (International Council of Air Shows, 2023a; UK Civil Aviation Authority, 2023a).

To qualify for these positions, candidates must meet the specific requirements of their respective countries and organizations. In the United States, candidates must meet the qualification requirements set by the Federal Aviation Administration (FAA), while candidates must then complete an air boss training course and pass a written and practical exam to receive their certification (International Council of Air Shows, 2023a). Similarly, in the United Kingdom, candidates must have extensive experience in aviation and risk management to become flying display directors. They must complete a training course that covers the regulations and procedures for air shows, pass a written exam, and demonstrate their ability to manage an air show safely and efficiently (UK Civil Aviation Authority, 2023a).

FIGURE 1.11 The author (Manolis Karachalios), serving as a flying display director, is stationed in the control tower during the Athens Flying Week Air Show at Tatoi airfield, Greece, in 2013. Photo by the author.

The International Council of Air Shows (ICAS) and the European Airshow Council (EAC) also recognize the importance of air bosses and flying display directors in air shows. ICAS provides education and training for air show professionals, including the Air Boss Academy (International Council of Air Shows, 2023b), which covers topics such as air traffic management, risk management, and emergency procedures. The EAC offers a Flying Display Director Seminar (European Airshow Council, 2023) that covers the regulations and procedures for air shows in Europe, including designing and coordinating aerial displays, managing air traffic, and ensuring the safety of all participants.

In Australia, the display coordinator is appointed by the display organizer to the Civil Aviation Safety Authority (Civil Aviation Safety Authority, 2022). The display coordinator is in charge of the actual flight program and is accountable for the overall airborne component and safety of the event. In addition to the display coordinator, a display organizer may appoint a small group of experienced individuals to serve as a display committee for larger displays. Additionally, a component of air shows in Australia is the ground control coordinator, who is accountable to the display organizer. They should have a substantial and verifiable aviation background commensurate with the planned event, which enables them to identify aviation ground-based hazards and their impact on persons and property during the event.

Another function to ensure safety in air shows is applied in the United Kingdom: the Flying Control Committee (FCC), which oversees air show safety and augments the duties of the flying display director (FDD). The UK Civil Aviation Authority (CAA) appoints the FCC for each air show, which is responsible for ensuring that the air show is conducted safely and complies with the UK CAA's regulations and standards (UK Civil Aviation Authority, 2023a). The FDD selects the FCC, which should include pilots with experience flying the Flying Display's aircraft and might be augmented by other air show experts.

Air bosses, flying display directors, display coordinators, and the FCC play a crucial role in ensuring the safety and success of air shows, and their importance was highlighted in the study's findings. Air bosses serve as risk management intermediaries in air shows, bridging the gap between the regulators and the air show performers (see Figure 1.12). As such, they are held accountable for the safe execution of an air show. Ultimately, air bosses set the tone of the air show during the display safety briefing, and they must maintain the rhythm of the air show choreography until the end of the event.

1.2.3 Air Show Regulators

In addition to the categories of air show performers listed previously, aviation authorities provide regulations and guidelines for air show performers to ensure the safety of themselves and the audience. The French civil aviation authority, Directorate General for Civil Aviation (DGAC), provides specific regulations for air show performers in France, including rules for formations and jet demonstrations (French Civil Aviation Authority, 2023). Similarly, the UK Civil Aviation Authority (UK CAA) provides safety and administrative requirements and guidance for flying displays and special events (UK Civil Aviation Authority, 2023b). In Brazil, the National Civil Aviation Agency (ANAC) provides regulations for air shows and aerial demonstrations (Agência Nacional de Aviação Civil [ANAC], 2023). ANAC's Regulation No. 17/2015 outlines the necessary procedures and requirements for conducting these events (Agência Nacional de Aviação Civil [ANAC], 2015). Then, the South African Civil Aviation Authority (SACAA) oversees regulations for air show performers in South Africa. The South African Special Air Events (SAE) Handbook is one of the documents forming the required SAE documentation set, which specifies the requirements and restrictions applicable to air displays and special events (South African Civil Aviation Authority, 2021). Then, in the United Arab Emirates (UAE), the General Civil Aviation Authority (GCAA) is the main regulatory body responsible for overseeing the safety and security of civil aviation in the country

FIGURE 1.12 The relationship amongst air show stakeholders.

(UAE General Civil Aviation Authority, 2023). The GCAA has published requirements and standards that air show organizers, participants, and display teams must follow to ensure the safety of all involved.

Air show regulators are responsible for ensuring that air shows are conducted safely and in compliance with national and international regulations. They play a crucial role in setting standards and guidelines that organizers and performers must follow to reduce the risk of accidents and injuries. Additionally, regulators can change air show procedures to address safety concerns and new technologies or equipment.

For example, after the tragic incident at the Shoreham Airshow in the UK in 2015, air show regulators reviewed safety procedures and made significant changes (UK Air Accidents Investigation Branch, 2017). The accident was a wake-up call for the air show industry, prompting a review of safety procedures and regulations. The UK Civil Aviation Authority (CAA) implemented stricter regulations, such as requiring more detailed risk assessments from air show organizers. These changes were made to mitigate the risks of accidents and improve safety for all involved.

Another example of regulators changing air show procedures is the Federal Aviation Administration (FAA) in the United States. In response to the COVID-19 pandemic, the FAA issued guidelines for air shows that included measures to reduce the risk of virus transmission. These measures included reducing the number of spectators and requiring performers and staff to follow social distancing and hygiene protocols. The FAA's changes aimed to protect public health while allowing air shows to continue safely.

In addition to responding to specific incidents or concerns, regulators can change air show procedures to incorporate new technologies or equipment. For instance, introducing unmanned aerial vehicles (UAVs) or drones has led to new regulations and guidelines from air show regulators. The FAA has developed guidelines for the safe operation of UAVs at air shows, including restrictions on their use during certain maneuvers or in certain areas.

In conclusion, air show regulators play a critical role in ensuring the safety of air shows. Air show regulators exist in many countries around the world, and even though their roles and responsibilities may vary depending on the region, they all align in setting standards and guidelines, conducting safety assessments, and making changes to air show procedures as needed so that they can mitigate the risks of accidents and injuries. As new technologies emerge and concerns arise, regulators will continue to play an essential role in ensuring that air shows are conducted safely and in compliance with regulations.

1.2.4 Air Show Stakeholders Association/Collaboration/Bonding

In addition to the national aviation regulatory agencies, other institutions, such as the International Council of Air Shows (ICAS), the European Airshow Council (EAC), the Fédération des Spectacles Aériens (FSA), and the British Air Display Association (BADA), provide valuable support to the air show industry with their knowledge, expertise, and industry best standards that may be more stringent than regulatory requirements. Furthermore, the Fédération Aéronautique Internationale (FAI), also known as the World Air Sports Federation, is an international organization that oversees air sports, including aerobatics, which are often featured in air shows. The FAI establishes rules and safety standards for various air sports and maintains a global network of national aviation organizations.

In recent decades, the international air show community has also made significant strides in the area of safety and risk management. The establishment of organizations such as the ICAS, the EAC, the FSA, the FAI, and the BADA has played a crucial role in promoting safety standards and best practices across the industry. These efforts have helped to ensure that air shows remain a thrilling yet secure celebration of human achievement.

1.2.5 Diversity and Inclusion in Air Shows

The air show community has come a long way in promoting diversity and inclusivity, with performers from diverse backgrounds, genders, and ethnicities taking center stage at events across the globe. This progress reflects the growing recognition of the importance of diversity in aviation and the need to inspire future generations of pilots and performers from all walks of life. As the world becomes increasingly interconnected, the air show community needs to continue fostering an environment that welcomes and celebrates diversity.

By showcasing the talents and achievements of performers from diverse backgrounds, air shows contribute to a more inclusive aviation community and inspire individuals who may have previously felt excluded or underrepresented. As we continue to break down barriers and promote inclusivity, the sky is truly the limit for what we can achieve in the world of air shows and beyond.

1.2.5.1 Historical Perspectives

There has been incremental progress in the area of diversity and inclusion within the international air show community from the early 1920s till date. Male pilots dominated the early days of air shows, but as the years progressed, more women and people of diverse backgrounds began taking to the skies. Pioneers like Bessie Coleman, the first African American woman to hold a pilot license, broke barriers and shattered stereotypes in the 1920s (Rich, 1993). Despite facing immense adversity, including racial and gender discrimination, Coleman's determination and talent inspired generations of aspiring pilots from diverse backgrounds. Another trailblazer in the world of air shows was Pancho Barnes, born as Florence Leontine Lowe. She was a pioneering aviator, stunt pilot, and founder of the first movie stunt pilots' union. Pancho Barnes set numerous aviation records, including breaking Amelia Earhart's air speed record in 1930, and her fearless attitude and skills paved the way for many female pilots that followed.

In Europe, one of the earliest female air show pilots was Marie Marvingt, a French aviator who was also an accomplished athlete, nurse, and journalist. She was the first woman to fly combat missions as a bomber pilot during World War I and participated in numerous air shows, showcasing her skills and encouraging other women to take up flying.

Currently, Patty Wagstaff and Svetlana Kapanina are indeed remarkable contemporary female air show performers who have made significant contributions to the world of aerobatics.

Patty Wagstaff, an American aerobatic pilot, is renowned for her incredible skill and precision in the air (see Figure 1.13). Born in 1951, she began her flying career in the early 1980s and quickly gained recognition for her talent. In addition to being a three-time U.S. National Aerobatic champion, Wagstaff has also won numerous international competitions and received several prestigious awards. She was inducted into the National Aviation Hall of Fame in 2004, further solidifying her legacy in the world of aviation. Today, she continues to inspire and train the next generation of pilots through her flight school and by performing at air shows worldwide.

Svetlana Kapanina, born in 1968, is another outstanding aerobatic pilot representing Russia on the global stage. With her exceptional skills and dedication, Kapanina has secured seven World Aerobatic Champion titles in the women's category, making her one of the most successful female pilots in the history of the sport. She has also won numerous European championships and other international awards, earning the respect and admiration of her peers and fans. Kapanina continues to compete and perform in air shows, further pushing the limits of what is possible in aerobatics and inspiring future generations of pilots.

Today, organizations like Women in Aviation International (WAI), the Organization of Black Aerospace Professionals (OBAP), and the National Gay Pilots Association (NGPA) provide strong advocacy and work tirelessly to promote diversity and inclusion within the industry. By providing scholarships, mentorship

FIGURE 1.13 Patty Wagstaff, flying her Extra 300. Copyright: Patty Wagstaff.

programs, and networking opportunities, these organizations are fostering an environment where everyone, regardless of their background or identity, can succeed and thrive in aviation careers such as air show performance.

Air shows have also become platforms for showcasing the talents of diverse pilots and performers. Events such as the Sisters of the Skies outreach programs and the Warriors Over the Wasatch Air and Space Show demonstrate the power of inclusion and the importance of representation in inspiring future generations of aviators. The Sisters of the Skies outreach programs, for example, highlight the achievements and talents of women pilots, showcasing their skills and breaking down gender stereotypes. This empowers and encourages more women to pursue careers in aviation, thus increasing diversity in the field. On the other hand, the Warriors Over the Wasatch Air and Space Show focus on celebrating the contributions of disabled veterans, who may have unique experiences and skills to share. By highlighting the achievements of these veterans, the event helps to raise awareness about the capabilities of disabled individuals and fosters a more inclusive environment in the world of aviation.

Another shining example of diversity and inclusion in air shows is the WeFly! Team, an Italian aerobatic display team made up of pilots with disabilities (see Figure 1.14). Founded in 2007, the team showcases the incredible skills and resilience of disabled pilots, challenging preconceived notions about their capabilities. The WeFly! Team's performances not only demonstrate the power of determination but also serve as a powerful symbol of inclusivity in the world of aviation.

Air shows today are a melting pot of diverse performers, showcasing talent from around the globe. Pilots from different countries, cultures, and backgrounds come together to share their passion for aviation, demonstrating that the sky truly has no borders. The increased visibility of diverse pilots in air shows has helped to foster a more inclusive environment, inspiring a new generation of aviators and enthusiasts from a wide variety of demographics.

FIGURE 1.14 The WeFly! Team performing in close formation at AFW 2013, Tatoi AFB. Copyright: Athens Flying Week (Photo by WeFly!).

1.2.6 Trailblazing Female Airs Show Performers

Numerous trailblazing female air show performers have left their mark on history, but two individuals, in particular, significantly impacted their generation: Bessie Coleman (1892–1926) and Maryse Bastié (1898–1952). Their remarkable accomplishments and tenacity inspired their contemporaries and paved the way for future generations of female aviators.

Bessie Coleman and Maryse Bastié proved that women could excel in male-dominated fields given the opportunity. Their accomplishment in aviation is a true testimony of their resilience and exceptional skills. These two aviatrixes have left a lasting legacy that continues to inspire countless numbers of women and marginalized people to be relentless in the pursuit of their dreams, even if it means challenging established conventions. Their stories highlighted in the next section show that perseverance, courage, and passion are required to overcome challenges in the quest for success.

1.2.6.1 Bessie Coleman: Breaking Barriers in the Sky

Bessie Coleman (1892–1926) was a pioneering American civil aviator who broke racial and gender barriers by becoming the first African American woman and Native American woman to hold a pilot's license (see Figure 1.15). Born in an era marked by discrimination and limited opportunities for women, especially women of color, Coleman's determination to achieve her dream of flying was nothing short of inspiring.

FIGURE 1.15 American pilot Bessie Coleman and her bi- plane circa 1922. Copyright: Michael Ochs Archives / Getty Images.

1.2.6.1.1 Early Life

Born in Texas on January 26, 1892, to a Native American father and African American mother, Bessie Coleman faced the double difficulties of racial and gender discrimination in early 20th-century America. Despite these challenges, Coleman was determined to pursue her dream of flying. Interestingly, her brother John's teasing, which included comparing her to French women who could fly, played a significant role in motivating her to seek a pilot's license.

However, Coleman's journey to becoming a pilot was far from easy. Due to her gender and race, American flight schools rejected her. Undeterred, Coleman sought the help of Robert Abbott, publisher of the well-known African American newspaper, the Chicago Defender. Abbott encouraged Coleman to save money, move to France, and pursue her pilot's license there. Taking his advice to heart, Coleman worked tirelessly to save funds and learn French, eventually moving to France in November 1920.

1.2.6.1.2 Aviation Achievements

After earning her pilot's license, Bessie Coleman returned to the United States and quickly gained fame as an air show performer, known as "Queen Bess." Her aerial stunts, which included daring loop-the-loops and barrel rolls, captivated audiences across the country. Coleman's skills as a barnstormer demonstrated her exceptional abilities as an aviator and set her apart from her contemporaries.

One of Coleman's most remarkable achievements was her first public air show on September 3, 1922, at Glenn Curtiss Field in Garden City, New York. Sponsored by the Chicago Defender, this event showcased Coleman's exceptional skills and catapulted her to national celebrity status. Her performances not only entertained audiences but also served as an inspiration for other aspiring pilots, particularly women and African Americans.

1.2.6.1.3 Legacy and Impact

Although Bessie Coleman's tragic death in 1926 cut her aviation career short, her impact on aviation history was profound. As the first African American woman and Native American woman to earn a pilot's license (National Air and Space Museum, 2021) and perform daring aerial stunts, she inspired future generations of African American and Native American pilots to break barriers in the field. Coleman's life and accomplishments demonstrated that with determination and hard work, individuals from all backgrounds could achieve success in the aviation industry.

Even after Bessie's demise, her accomplishment inspired the formation of multiple Bessie Coleman Aero Clubs, the organization of the first all African American Air show in 1931, and an annual flyover of her grave by African American pilots. Additionally, her name appeared on buildings in Harlem, further solidifying her status as a role model and trailblazer in aviation. Coleman gained fame outside the black community over time. In the afterword to "Queen Bess: Daredevil Aviator," Mae Jemison, the first African-American woman in space in 1992, (Slotnik, 2019) writes that she was "embarrassed and saddened that I did not learn of her until my spaceflight beckoned on the horizon." Jemison carried a photo of Coleman into orbit, traveling higher than Coleman had dreamt.

As a trailblazer in the aviation industry, Bessie Coleman's accomplishments continue to inspire and motivate individuals from diverse backgrounds to pursue their dreams. Her remarkable journey from humble beginnings to becoming an aviation icon serves as a testament to her resilience and tenacity. Coleman's story is a powerful reminder that with passion and determination, one can overcome seemingly insurmountable obstacles.

1.2.6.2 Maryse Bastié: A Trailblazer in the Skies

Maryse Bastié (1898–1952) was a pioneering French aviator who shattered gender barriers and set multiple records during her illustrious career (Fédération Aéronautique Internationale [FAI], 2019). Bastié's achievements were awe-inspiring, considering the societal norms and expectations of women during her time. She demonstrated exceptional skill, courage, and determination, inspiring a new generation of female pilots, such as Jacqueline Auriol and Hélène Boucher, to continue breaking barriers in aviation.

1.2.6.2.1 Early Life

Maryse Bastié was born Marie-Louise Bombec on February 27, 1898. She faced considerable hardships throughout her life, starting with her father's death when she was just 11 years old. As a result, Bastié was forced to work in a factory to support her family.

Despite her tragedies, Bastié was determined to rise above her circumstances. In 1921, she met and married her second husband, the World War I pilot Leon Bathiat. It was through him that Bastié discovered her passion for aviation.

1.2.6.2.2 Aviation Achievements

Maryse Bastié's aviation career began in 1925 when she obtained her pilot's license. This was a remarkable accomplishment, considering that few women were allowed to fly at the time. Bastié quickly gained recognition for her skills, and in 1927, she set the women's world record for the longest solo flight, covering a distance of 1,243 miles.

Bastié continued to break records throughout her career. In 1929, she set the women's world record for endurance, flying for 26 hours and 47 minutes. In 1931, she broke the women's world speed record, reaching an average speed of 126.48 miles per hour (Pujol, 2014). Her most significant achievement came in 1936 when she established a new world record for the longest nonstop flight by a woman, flying 2,976 miles from Paris to Moscow in 37 hours and 55 minutes.

Bastié received numerous awards for her accomplishments, including the Harmon Trophy in 1931 and the Légion d'honneur in 1932. In 1935, she was promoted to the rank of captain in the French Air Force.

1.2.6.2.3 Legacy and Impact

Maryse Bastié's groundbreaking achievements in aviation had a profound impact on women's involvement in the field. She served as a role model for other female aviators, proving that women could excel in a male-dominated industry. A new generation of female pilots, including Jacqueline Auriol and Hélène Boucher, were motivated by her success to keep breaking down boundaries in aviation.

Bastié's legacy is preserved through various initiatives that honor her memory. In 1955, the Maryse Bastié Foundation was established to provide scholarships for young women pursuing careers in aviation. Additionally, numerous schools, streets, and airports in France bear her name, a testament to her lasting impact.

Maryse Bastié was a remarkable woman who defied societal norms and overcame adversity to achieve extraordinary success in the world of aviation. Her tenacity and determination enabled her to break multiple records and establish herself as one of the most accomplished female pilots of her time. Bastié's legacy lives on through the generations of women she inspired to challenge convention and pursue careers in aviation. Her story serves as a powerful reminder of the importance of perseverance, courage, and passion when confronted with challenges.

1.2.7 All African American Air Show Performers

African American air show performers have played a critical role in shaping the landscape of aviation history by breaking racial barriers and inspiring generations of aspiring pilots. These trailblazers, such as The Five Blackbirds, defied the odds and overcame discrimination to showcase their talents in the skies. Their accomplishments captivated audiences and demonstrated that skill, determination, and passion transcended race. These exceptional individuals paved the way for greater diversity and inclusion within the aviation industry, fostering a sense of unity and camaraderie among pilots of all backgrounds. Today, the legacy of all African American air show performers continues to inspire and motivate aspiring aviators, reminding them that the skies are a boundless realm for those who dare to soar above the challenges of race and societal expectations.

1.2.7.1 The Five Blackbirds

In 1934, a group of African Americans, mostly veterans of World War I, formed a group called "The Five Blackbirds" to promote aviation in the African American communities (Aircraft Owners and Pilots Association [AOPA], 2009). The group also formed flying clubs and trained pilots and mechanics, and promoted an interest in aviation through publications, lectures, and "air circuses." The "air circuses" of the group entailed aerial acrobatics, rolls, turns, spins, and ribbon cutting, to name a few.

40,000 people attended the inaugural all-black Air Circus to see the world's first all-black aerobatic team perform. Julian went off as a passenger for the first act, performing a triple parachute jump. After the first canopy opened, he'd cut it away, freefall, then open the second canopy, which he'd also cut away, freefall, then open and touchdown with the third. Julian appears to have lost his nerve and plummeted beneath the first parachute, but then he climbed his aircraft to 6,000 ft.

The Blackbirds took to the skies after Julian. They flew through the crowd in a trail formation, while performing a sequence of breaking-off and rejoin flypasts, while the crowd was thrilled with these unique aerial acts.

Another achievement of members of the group was the long-distance flight of C. Alfred Anderson and Dr. Albert E. Forsythe, which consolidated the skills and airmanship of these individuals and appealed to a need for recognition and equality in aviation. Concurrently, William J. Powell set up a Bessie Coleman Aero Club in Los Angeles and authored an inspiring book, *Black Wings*, which encouraged Black youth to choose careers in aviation. The legacy and solid foundation laid by groups such as The Five Blackbirds continue to inspire and motivate aspiring aviators, especially those of color, and remind them of the boundless realm for those that dare to surmount the barriers of racial and societal expectations.

1.2.8 Global Perspectives

Air shows today are a melting pot of diverse performers, showcasing talent from around the globe. Pilots from different countries, cultures, and backgrounds come together to share their passion for aviation, demonstrating that the sky has no borders. A more inclusive atmosphere has been cultivated by the greater visibility of various pilots in air shows, which has inspired a new generation of aviators and followers from all backgrounds.

1.2.8.1 North America

1.2.8.1.1 United States

The U.S. air show circuit has indeed experienced a growth in diverse performers in recent years, with groups and individuals from various backgrounds showcasing their talents and contributing to the vibrant world of aviation. The inclusion of more diverse performers has not only added to the excitement and variety of air shows but also serves to inspire the next generation of pilots and aviation enthusiasts from all walks of life.

The all-female Misty Blues Skydiving Team is an excellent example of this increased diversity (MistyBlues, 2023). Founded in the 1980s, the team has evolved over the years and is now composed of a dynamic group of women who perform thrilling skydiving routines at air shows across the United States. The Misty Blues represent a positive shift in the male-dominated world of aviation, as they challenge stereotypes and demonstrate the capabilities of female pilots and skydivers.

Another prominent figure in the diverse air show circuit is the African American pilot Anthony Oshinuga (Oshinuga, 2023). A skilled pilot and passionate aviation enthusiast, Oshinuga (see Figure 1.16) has made a name for himself in both aerobatics and vintage racing events. In addition to his impressive flying skills, Oshinuga is also a motivational speaker and advocate for diversity in aviation. He regularly shares his journey and experiences with others, aiming to inspire young people, especially those from underrepresented backgrounds, to pursue careers in aviation.

FIGURE 1.16 Anthony Oshinuga is performing a low-level pass with his Pitts Special. Copyright: Anthony Oshinuga.

1.2.8.1.2 Canada

Originally from Ukraine, Anna Serbinenko is a Canadian aerobatic pilot known as the "Sky Dancer" (McIntyre, 2017). She performs ballet-like routines in her Super Decathlon aircraft, captivating audiences with her graceful aerial displays. Serbinenko is also a certified flight instructor, and she is passionate about empowering women in aviation.

1.2.8.2 Europe

Throughout Europe, various air shows have made concerted efforts to promote diversity and inclusivity. Events such as the Malta International Air Show and the Finnish Air Force's Midnight Sun Air Show have welcomed performers from various countries and backgrounds, such as the Royal Jordanian Falcons, fostering a sense of global unity and camaraderie within the aviation community.

1.2.8.2.1 United Kingdom

In the United Kingdom, the Royal International Air Tattoo (RIAT) is one of the world's largest and most prestigious air shows. The event has historically showcased a variety of international performers, reflecting the diverse nature of the global aviation community. Additionally, the UK has seen a steady increase in female air show performers, such as Kirsty Murphy, the first female Red Arrows pilot and a member of the Blades Aerobatic Team (RIAT, 2020).

1.2.8.2.2 France

France is home to the renowned Patrouille de France, the national aerobatic demonstration team, which has made strides in promoting diversity within its ranks. In 2009, Capt. Virginie Guyot became the first woman to join the Patrouille de France as a full-time pilot (Bremner, 2008). Additionally, France hosts the Paris Air Show, a global event that has increasingly showcased performers from diverse backgrounds and nationalities.

1.2.8.3 Africa

In Africa, the Aero South Africa event is a prime example of an air show that has embraced diversity and inclusivity, featuring performers from across the continent and beyond. One notable African air show performer is Mandisa Mfeka (Matshili, 2019), South Africa's first Black female combat pilot, who has inspired countless young women to pursue careers in aviation.

1.2.8.4 Australia

Australia has also seen a rise in diverse air show performers in recent years. For example, the Avalon Airshow features performers from various backgrounds and nationalities, showcasing the talents of both local and international pilots. Australian aerobatic team Red Baron (2023) is another example of performers who actively promote diversity within the aviation community, as the team includes pilots from a variety of backgrounds and skill sets.

1.3 WHERE

1.3.1 Soaring Beyond Borders: The Inspiring Evolution of Air Shows Internationally

Air shows have long captivated the hearts and minds of millions of spectators around the globe. These enthralling spectacles showcase the boundless potential of human innovation and the indomitable spirit of those who dare to defy gravity.

The birth of air shows can be traced back to the early 20th century when intrepid aviators took to the skies in fragile, wood, and fabric aircraft to demonstrate their newfound mastery of flight. The first international air show, the Grande Semaine d'Aviation de la Champagne, took place in France in 1909, attracting pioneering pilots from across Europe and the United States. This historic event marked the beginning of a captivating journey that would see air shows evolve into the breathtaking displays of skill, speed, and technology we witness today.

One of the most notable advancements in the evolution of air shows has been the development of new aircraft technologies and aerobatic techniques. From the daring feats of early pilots such as Lincoln Beachey, who wowed crowds with his daring loop-the-loop maneuvers, to the jaw-dropping performances of modern-day aerial artists like Kent Pietsch and Rob Holland, air show performers have continually pushed the boundaries of what is possible in the realm of aerial acrobatics.

The inspiring evolution of air shows internationally stands as a testament to the resilience, adaptability, and determination of the pilots, engineers, and organizers who have dedicated their lives to this enthralling pursuit. As we continue to witness remarkable advancements in technology, safety, and diversity, we can only imagine what the future holds for the captivating world of air shows.

The growing international cooperation and exchange of ideas among air show performers and organizers have led to the emergence of breathtaking performances that transcend national boundaries. Iconic teams such as the United States Air Force Thunderbirds, the Royal Air Force Red Arrows, and the Patrouille de France have become symbols of pride and excellence, showcasing the collaborative spirit and shared passion for aviation that unites nations.

In recent years, the international air show community has also embraced environmental sustainability, with events such as the Farnborough International Airshow and the Paris Air Show showcasing cutting-edge technologies and innovations aimed at reducing the environmental impact of aviation. This commitment to sustainability highlights the forward-thinking nature of the air show industry and its dedication to preserving our planet for future generations.

The continued growth and popularity of air shows internationally demonstrate the enduring fascination with flight and the power of these aerial spectacles to inspire and unite people across the globe.

1.3.2 Today's Types of Air Show Events

There are various types of aerial events within the air show industry, each designed to captivate audiences and display remarkable human achievements in aviation. These events range from ceremonial flypasts to impressive military demonstrations, all sharing a common goal of entertaining spectators and celebrating the wonders of flight. Examples of such events include the following.

1.3.2.1 Flying Displays

Flying displays are one of the most common types of air show events, and they involve a variety of aircraft performing aerobatic maneuvers and flying in formation. These displays are often accompanied by music and commentary, aiming to showcase pilots' skill and precision. Flying displays can also include flypasts, in which aircraft fly over the crowd at low altitudes.

Formation flying is often featured in aerial displays, showcasing pilots flying in close proximity while executing precise maneuvers. This type of flying demands exceptional skill and coordination, as it highlights the capabilities of the aircraft involved. Additionally, formation flying can be employed to craft artistic formations in the sky, captivating audiences with its unique visual appeal.

1.3.2.2 Static Displays

Static displays serve as a popular component of air shows, providing spectators with the opportunity to closely examine various aircraft stationed on the ground. These exhibitions often showcase military planes, helicopters, and vintage aircraft, offering visitors a chance to explore the history and technological advancements of aviation. Furthermore, static displays can be utilized for recruitment and promotional purposes, engaging the public with the fascinating world of flight.

1.3.2.3 Night Shows

Night shows are a more recent addition to the air show circuit, and they involve flying displays that take place after dark, often accompanied by fireworks and other special effects. Moreover, night shows are designed to create a unique and memorable experience for spectators, and they often feature aerobatic displays lit up with colored lights.

1.3.2.4 Warbird Shows

Warbird shows, a specialized type of air show event, concentrate on showcasing vintage military aircraft. These aircraft frequently fly in formation, engaging in simulated dogfights and other military tactics. These events captivate aviation enthusiasts and history aficionados alike, providing a window into the storied past of military aviation. Additionally, warbird shows serve as a tribute to veterans and a means of fostering patriotism among attendees.

1.3.2.5 Military Demonstrations

Military demonstrations represent a distinct category of air show events featuring military aircraft executing a variety of maneuvers and tactics. These demonstrations serve as valuable tools for recruitment and marketing, as they enable the military to display its advanced technology and capabilities to potential recruits and the wider public. Furthermore, military demonstrations foster patriotism and national pride among attendees.

1.3.2.6 Air Races

Air races form an exciting component of air show events, with pilots pitting themselves against one another in a timed race along a designated aerial course. These races aim to exhibit the speed and agility of various aircraft while also offering manufacturers an opportunity to showcase their aircraft's capabilities. Additionally, air races can contribute to promoting tourism and fostering economic development in the host regions.

1.3.2.7 Aerobatic Competitions

Aerobatic competitions are a subset of air show events that feature talented pilots performing various aerobatic maneuvers such as loops, rolls, and dives. These competitions aim to promote aerobatics while also highlighting the expertise and precision of participating pilots. Aerobatic competitions also serve as a platform for marketing and sponsorship opportunities, increasing their appeal and reach.

1.3.2.8 Fly-ins

Often hosted at smaller airports or airstrips, fly-ins are a less formal form of an air show where pilots can display their aircraft to aviation enthusiasts. These gatherings give pilots a fantastic opportunity to network with others who share similar interests. Moreover, fly-ins provide the opportunity to learn about new technologies and aircraft designs. They can also function as community forums for promoting aviation safety and education.

Overall, air show events come in many different forms, and each type offers a unique experience for spectators and participants alike. Whether it's watching skilled pilots perform aerobatic maneuvers, exploring the history of aviation through static displays, or socializing with other aviation enthusiasts at a fly-in, there is something for everyone at an air show.

1.4 WHAT

1.4.1 Risk Management: From Daredevils to Professionals

Risk management is an essential aspect of air show safety, as it enables pilots and organizers to identify and mitigate potential hazards, allowing them to push the boundaries of flight while preserving the well-being of all involved. The remarkable evolution of risk management techniques and strategies, from the early days of aviation to the present, stands as a testament to the air show community's resilience, adaptability, and determination. In writing about risk management, we seek to inspire readers with the ingenuity and resourcefulness of the individuals who have dedicated their lives to ensuring that air shows remain a thrilling yet secure celebration of human achievement.

In the early 20th century, air shows were primarily about showcasing the daring feats of fearless aviators. Pilots like Linkoln Beachy in the United States, Emmanouil Argyropoulos[1] in Greece, and Maryse Bastié in France captured the public's imagination and inspired a generation of aviation enthusiasts. However, the high-risk nature of these early exhibitions led to numerous accidents and fatalities, prompting the need for more robust safety protocols and risk management techniques.

1.4.2 Current Approach to Risk Management

Over time, air shows have become increasingly safer as organizers, performers, and regulators work together to ensure the safety of both pilots and spectators. The engagement of organizations such as the International Council of Air Shows (ICAS) and the Federal Aviation Administration (FAA) played a crucial role in improving safety at air shows. They have set guidelines for minimum altitude restrictions, spectator safety zones, and pilot and air boss qualifications, which have significantly reduced the risks associated with air show performances.

Furthermore, aircraft design and engineering advances have also played a vital role in risk management. Modern aircraft are now equipped with sophisticated safety systems, including reliable engines, advanced flight controls, and instrumentation, which allow pilots to execute complex maneuvers with greater precision and confidence. One notable development in risk management is the use of sophisticated simulation software, allowing pilots to practice complex maneuvers and perfect their routines before taking them to the skies.

1.5 WHY

1.5.1 Evolution of Psychological Factors Affecting Air Show Performers

The psychological factors that shape the minds of air show performers are often overlooked. Still, they are crucial to understanding the incredible feats of these pilots and the challenges they face in their pursuit of excellence. Throughout the history of air shows, psychological factors and biases have played a significant role in shaping the decisions and actions of air show performers. By examining historical examples and comparing past and present situations, we can better understand how these factors have evolved.

During the early 20th century, aviation pioneers like Glenn Curtiss and Louis Paulhan demonstrated daring stunts and pushed the boundaries of human flight. The psychological factor of overconfidence may have influenced their decisions, leading to an underestimation of the risks involved. For instance, French pilot Adolphe Pégoud made history in 1913 (Reichhardt, 2010) by becoming the first pilot to perform a loop, but his boldness eventually led to his death during World War I.

Most of the early air show pilots, such as the pre–World War I spud flyers and the post-WW I barnstormers, entertained their audiences with awe-inspiring feats. As stated earlier, most of these pilots were predominantly male and willingly faced extreme safety risks, performing stunts that pushed their aircraft beyond their design capabilities leading to frequent and numerous fatalities (Onkst, n.d.; Prendergast, 1980). Tragically, many of these air show performers lost their lives in pursuit of these exhilarating displays (Baker, 1994; Casey, 1981; Friedlander & Gurney, 1973; Jablonski, 1980; Onkst, n.d.). Based on the high accident and fatality rates, it is plausible that apart from the poor design of the aircraft, human factors also contributed to these mishaps. In an era of hazardous attitudes such as machismo and invulnerability were desired attributes of great air show performers, there is a likelihood that these attitudes could have resulted in poor risk perceptions and unsafe actions (Federal Aviation Administration, 2016). Other psychological risk factors that could have affected these performers were the illusion of control, confirmation biases, and plan continuation biases, which can lead to unsafe behaviors during air show activities (Dekker, 2017; Nickerson, 1998; Thompson, 1999; You et al., 2013). These psychological risk elements have been suggested to influence highly skilled performers' decision-making processes (Wang & Zhang, 2020). Some of the early air show performers who excelled in such high-risk and often complex tasks may have been vulnerable to these psychological risk factors and pushed the limits of their flying skills despite being fatigued or overwhelmed with stress leading to fatal outcomes.

In the post–World War II era, aerobatic demonstration teams, such as the United States Air Force Thunderbirds and the United States Navy Blue Angels, emerged, showcasing their teamwork and professionalism. However, the psychological bias of groupthink (Janis, 1972) could have potentially limited dissenting opinions about the risks involved in their routines. During the 1950s, the Blue Angels faced a series of accidents, prompting a renewed emphasis on safety and risk management. While specific information about the causes of these accidents remains limited, it is plausible that some of the contributing factors were related to psychological elements. Common psychological factors that may have played a role in these accidents include cognitive biases, stress, fatigue, and breakdowns in teamwork and communication.

The late 20th century witnessed significant changes in the air show industry as performers like Julie Clark broke gender barriers and pushed the limits of aerobatic performances. However, these trailblazers may have encountered the psychological factor of stereotype threat, where the fear of confirming negative stereotypes about their gender or social group could have impacted their performance (Steele, 1997). Stereotype threat can have deleterious effects on performance by increasing anxiety, reducing working memory capacity, and shifting attention away from the task at hand (Schmader et al., 2008). In the context of air shows, these effects could have potentially hindered pilots' abilities to focus on complex flying maneuvers, make quick decisions, and maintain situational awareness. This, in turn, could have increased the risk of errors and accidents, compromising the safe execution of aerobatic performances.

In the 21st century, air shows have become more diverse and innovative, with performers such as the Red Bull Air Race pilots and the Breitling Jet Team showcasing impressive skills and cutting-edge aircraft technology. However, the increased demand for spectacular performances may lead to the psychological bias of self-serving attribution, where performers attribute their success to their skills and training while dismissing the role of luck or external factors. This bias could lead to a false sense of security, potentially exacerbating the risks involved in their performances.

Since 2010, the air show industry has placed a strong emphasis on safety. The tragic Shoreham accident in 2015, where the jet crashed into bystander spectators, killing 11 people and injuring 13 others, including the pilot, led to increased scrutiny and the implementation of new safety measures, which focused on enhancing the training, competency, and decision-making processes of the pilots and air show organizers. The UK Civil Aviation Authority (CAA) introduced new rules for display pilots, which included more stringent training requirements. This aimed to ensure that pilots are better prepared to handle challenging situations and make safe decisions during air displays (Air Accidents Investigation Branch, 2017). Moreover, as a result of the more stringent rules, air show organizers are required to conduct thorough risk assessments that take into account not only technical factors but also human factors, such as pilot experience, fatigue, and decision-making.

Additionally, the advent of new technologies, such as drones and virtual reality, has led to a new set of psychological factors and challenges for air show performers. For example, the Drone Racing League (DRL) has emerged as a popular form of air racing. Drone pilots may experience unique psychological factors, such as decision fatigue and information overload, as they manage complex flight controls and navigate intricate racecourses.

Historically, air shows have been powerful platforms for diplomacy and showcasing national pride, as evidenced by the United States' participation in international air shows like the Paris Air Show and the Farnborough International Airshow. Performers representing their countries in these events may experience increased psychological pressure to succeed, as well as potential biases stemming from nationalistic pride.

These historical examples could give a taste of the evolution of psychological factors and biases affecting air show performers throughout the years. From the overconfidence of early aviation pioneers to the various biases experienced by barnstormers, aerobatic teams, and trailblazing performers, these factors have played a significant role in shaping the air show industry. As air shows continue to evolve and innovate in the 21st century, understanding and addressing these psychological factors and biases will remain crucial for the ongoing success and safety of air shows and their performers.

1.5.1.1 Psychological Factors for Air Show Performers in the Age of Social Media

In recent years, the rise of social media has introduced new psychological factors and pressures that impact air show performers. The widespread use of social media platforms has significantly changed the way performers engage with their audiences, promote their shows, and receive feedback on their performances.

One psychological factor associated with social media is the need for validation and social approval. Air show performers, like many individuals and professionals, can become preoccupied with the number of likes, shares, and followers they receive on their social media profiles. This need for validation may push performers to constantly seek new, exciting, and potentially risky maneuvers to generate more attention and engagement from their online audience.

Social comparison is another psychological factor that social media can intensify. Performers may compare their skills, performances, and public recognition with other air show performers, leading to feelings of inadequacy or pressure to outdo their peers. This competitive mindset could contribute to increased risk-taking in an attempt to set themselves apart from other performers and gain a competitive edge.

The fear of missing out (FOMO) is another psychological factor that may affect air show performers in the age of social media. FOMO refers to the anxiety that individuals may experience when they perceive that they are missing out on rewarding experiences others enjoy (Przybylski et al., 2013). For air show performers, FOMO may manifest in fear of being left behind in terms of innovation or not participating in high-profile events. This fear may drive performers to push themselves harder or take on additional commitments, potentially leading to burnout or increased risk-taking.

Lastly, the immediacy and visibility of social media can amplify the impact of both positive and negative feedback on-air show performers. This heightened sensitivity to feedback may affect performers' mental well-being and their approach to their craft. While praise and support from fans can boost performers' self-esteem and motivation, negative comments or criticism can lead to self-doubt and increased pressure to improve.

In conclusion, the rise of social media has introduced new psychological factors and pressures that impact air show performers in various ways. The need for validation, social comparison, the fear of missing out, and the amplified impact of feedback can shape performers' decision-making, risk-taking, and overall well-being. Performers need to be aware of these factors and develop healthy strategies for managing their online presence and engagement to maintain a balance between their professional and personal lives.

1.5.2 Air Show Performers' Mindset Over the Years

Air show performers' mindset has also changed over time. Several historical examples illustrate the significant changes in attitudes, motivations, and philosophies that have shaped the air show industry and can be used to trace this evolution.

In the early 20th century, air shows were a novel spectacle, with performers focused on pushing the boundaries of aviation and showcasing their daring stunts. The 1909 Reims Air Meet in France, the world's first air show, exemplified the spirit of experimentation and risk-taking that defined the early days of aviation. Then one of the most famous examples from this period is the 1910 Dominguez Air Meet in Los Angeles, the first major air show in the United States. The event featured groundbreaking demonstrations by early aviation pioneers such as Glenn Curtiss and Louis Paulhan.

These early air show performers were driven by the desire to prove the viability of human flight and to inspire a sense of wonder and awe in the public. Their mindset was characterized by resilience and determination as they faced numerous challenges and failures in their pursuit of flight. Limited understanding of aerodynamics meant they had to rely on trial and error to develop aircraft that could successfully take off, fly, and land. Many early designs failed to achieve sustained flight, leading to setbacks and crashes. Early aircraft were made from basic materials like wood, fabric, and wire, which provided limited strength, durability, and control, making flying a risky endeavor. Engine technology was not as reliable or powerful as modern engines, causing crashes and setbacks due to overheating and mechanical failures.

The lack of standardized controls and training made flying even more dangerous, and many accidents occurred as pilots struggled to learn how to control their machines. Limited funding and resources hindered progress, often forcing pioneers to abandon or delay projects. Public skepticism about the viability of human flight made it difficult for pioneers to garner support and funding. Harsh weather conditions and the absence of weather forecasting technology added to the challenges faced by early aviators. As aviation became more popular, governments introduced regulations that could hinder progress or limit the activities of early aviators, such as restrictions on flying in overpopulated areas.

During the interwar period between World War I and II, air shows became increasingly competitive and commercialized. Performers such as the "Barnstormers" traveled across the United States, offering aerial displays and rides to the public.

Performers such as Lincoln Beachey, known as the "Master Birdman," epitomized this mindset, pushing their aircraft's limits and physical capabilities to captivate audiences. This period also saw the rise of air racing, exemplified by the Cleveland National Air Races, which began in 1929, with a mindset characterized by the performers to entice audiences to put themselves at risk by breaking airspeed records, even if they were exceeding the structural limits of their aircraft. Glory-hungry times!

The post–World War II era marked a shift in the mindset of air show performers, with a focus on precision and teamwork. The United States Navy Blue Angels (Stone, 2021; U.S. Navy, 2023; Veronico, 2005) and the United States Air Force Thunderbirds (Neubeck, 2020; U.S. Air Force, 2023), founded in 1946 and 1953, respectively, exemplify this shift. These teams emphasized mental and physical preparation, maintaining strict training regimens and high-performance standards. That focus on teamwork and professionalism in the teams was a reflection of the broader military values of the time and served as a tool for recruitment and public relations. However, it needs to be highlighted that all military aerobatic teams exemplify this mindset, which emphasizes the significance of trust and camaraderie among team members.

In the late 20th century, air show performers began to incorporate more advanced aircraft and technologies into their performances. The introduction of highly maneuverable aircraft using fly-by-wire flight controls, like the F-16 Fighting Falcon and the F-18 Hornet, allowed for more dynamic and powerful displays, including vertical climbs and high-speed passes. These technological advancements expanded the range of possible maneuvers and challenged performers to continuously innovate their routines.

In the 21st century, the mindset of air show performers has continued to evolve with a greater emphasis on safety, sustainability, and inclusivity. Following high-profile accidents in the past, performers and organizers have become more diligent in implementing strict safety measures and guidelines.

While early aviators were often driven by a desire for fame and fortune, contemporary pilots are more focused on mastery, teamwork, and the pursuit of excellence. This change in attitude has made air shows less about individual showmanship and more about the collective artistry of skilled pilots working in harmony.

Modern air show performers understand the importance of mental preparation and discipline. Many rely on visualization, meditation, and stress management techniques to maintain their focus and composure during high-pressure performances. This psychological evolution has led to even more precise and awe-inspiring aerial displays, showcasing the incredible feats that humans can achieve when mind and machine unite.

1.6 WHEN

Air shows have been a captivating spectacle since their inception in the early 20th century, bringing together people from all walks of life to celebrate the beauty, power, and grace of flight.

From their humble beginnings to the modern-day spectacles that attract millions of spectators worldwide, air shows have consistently pushed the limits of human achievement and aviation technology. The legacy of air shows is one of innovation, courage, and dedication. Over the years, air shows have evolved to become an emblem of unity, diversity, and inclusion while constantly adapting to new safety and risk management measures.

1.6.1 Early Air Shows: A Public Display of Novel Aircraft Capabilities and Pilot Skills

1.6.1.1 Promotion of Early Air Shows: Utilitarian Reasons or Business Venture?

The groundbreaking flight of the Wright brothers' first heavier-than-air aircraft in December 1903 ignited passionate debates about the potential and practicality of airplanes. Some envisioned their use for utilitarian purposes, while others championed air shows as a means to captivate the public's interest in flight and provide economic value to aircraft builders and operators (Baker, 1994; Onkst, n.d.; Prendergast, 1980). Visionary plane builders like the Wright brothers and Glenn Curtiss recognized the public's fascination with airplanes and assembled teams of exhibition fliers who traveled across the country showcasing their remarkable creations (Whitehouse, 1965).

These early air show aviators engaged in thrilling competitions, pushing the boundaries of aerial stunts to determine who could fly the fastest, highest, and farthest (Friedlander & Gurney, 1973). These daring individuals tested the limits of aircraft design during a time when many engineers were grappling with fundamental aeronautical challenges (Harris, 1991; Prendergast, 1980).

1.6.1.2 Reims Air Meet: A New Paradigm in Air Show Performance

In August 1909, following Louis Blériot's historic crossing of the English Channel, the Reims Air Meet was organized to promote air shows. Capitalizing on the widespread media coverage and public interest in Blériot's feat, officials from the city of Reims transformed an open plain outside the city into a sprawling airfield and miniature metropolis to accommodate the anticipated crowds. Barber and beauty shops, telephone and telegraph offices, and a massive grandstand with a 600-seat restaurant overlooking the airfield were built for the occasion. From August 22 to 29, 1909, 22 of the world's top aviators convened to compete in the first organized international air meet.

As the inaugural event of its kind, the meeting captured the attention of numerous political and military leaders as well as the general public. Officially dubbed Le Grande Semaine D'Aviation de la Champagne (The Champagne Region's Great Aviation Week), the Reims Air Meet hosted a variety of prestigious contests, including those for the best flights in distance, altitude, and speed. Attractive cash prizes and impressive trophies enticed competitors to set new records in nearly every category. Throughout the week, spectators witnessed a gamut of emotions, from pure elation when their heroes triumphed to utter dismay when their favorites crashed. In essence, the Reims Air Show played a pivotal role in establishing aerial competitions as a premier form of entertainment in the early 20th century.

Among the 22 aviators who arrived in Reims to compete, all but two were Frenchmen. Notable French pilots included Louis Blériot, Hubert Latham, Henri Farman, and Eugene Lefebvre. The only foreign competitors were George Cockburn, a Scot, and Glenn Curtiss, an American. During the meet, Lefebvre demonstrated a series of daring aerial acrobatics, quickly earning a reputation as a daredevil. For the majority of attendees, the Reims Air Meet demonstrated that aviation competitions were not only tremendously exciting forms of entertainment but also showcased the economic and technological potential of aviation.

1.6.1.3 Chronicles of American Skies: A Historical Overview of U.S. Air Shows

Following the success of the Reims Air Meet, the United States hosted three major air shows in 1910, significantly shaping the future of American aviation. These events took place in Los Angeles, Boston, and New York, drawing large crowds eager to witness their first real aircraft. Pilots set new records at each meeting and showcased dazzling aerial stunts. As some scholars observe, these pioneering American air shows cultivated a sense of "air awareness" among attendees, making them appreciate not only the entertainment value of airplanes but also their practical potential. Notably, the 1910 U.S. air meets inspired numerous aspiring pilots, many of whom would become influential figures in the early exhibition era of aviation.

Official exhibition teams, often supported by renowned names like Wright and Curtiss, were formed to participate in competitions and air shows worldwide. The risk and excitement of these air shows attracted audiences despite the alarmingly high rate of accidents and fatalities among performers. Enthusiasm for air show displays surged, sparking public imagination for daredevil stunts in the air, with some air shows on the heavily populated U.S. East Coast drawing hundreds of thousands of spectators. As previously highlighted, the safety record of these air shows was abysmal, and historian David Onkst estimates that about 90% of exhibition pilots during this period perished in flight, a situation exacerbated by the onset of World War I, which further increased pilot fatalities. Although the war dampened the air show atmosphere's more lighthearted aspects, it paved the way for exponential growth in American aviation after the conflict.

1.6.1.4 Barnstorming Bonanza: The Golden Age of High-Flying Entertainment

In the aftermath of World War I and throughout the 1920s, barnstorming emerged as an immensely popular form of entertainment. Stunt pilots and aerialists, known as "barnstormers," performed a wide array of daring feats and tricks with airplanes. This era marked the first significant chapter in the history of civil aviation. For many pilots and stunt performers, barnstorming offered an exhilarating livelihood and an opportunity to showcase their creativity and showmanship.

The surplus of pilots and training aircraft in the 1920s set the stage for a golden era of American air shows, primarily small, informal "barnstorming" events. With the government selling surplus Curtiss "Jenny" trainers for a few hundred dollars, pilots could easily purchase one and quickly recover their investment by offering rides at small towns nationwide for as little as $1. Barnstorming acts often featured daring spins, dives, loops, and barrel rolls. With minimal regulation and ample open farmland for staging shows, barnstorming swept across America during the 1920s, drawing in pilots who would later leave lasting marks on aviation history.

Typically, a pilot or team of pilots would fly over a small rural town to attract attention, then land at a local farm (hence "barnstorming") and negotiate with the farmer to use a field as a temporary runway for staging an air show and offering airplane rides. After securing a location, pilots would fly back over the town, dropping handbills advertising airplane rides and aerial stunts for a small fee. The arrival of a barnstormer or aerial troupe in a rural town was often considered a spontaneous holiday, with locals flocking to buy plane rides and watch the show.

While many barnstormers operated independently or in small teams, some organized large "flying circuses" with multiple planes and performers. One of the most famous and well-traveled acts was the Ivan Gates Flying Circus, which toured nearly every state and extensively internationally. The Gates Flying Circus was known for popularizing the one-dollar joy ride and launching the careers of many renowned pilots.

Barnstorming flourished in North America during the first half of the 1920s but declined in the late 1920s due to new safety regulations and the cessation of surplus Jenny sales. These factors, combined with restrictions on aerial stunts and challenges in maintaining aging aircraft, led to the eventual demise of barnstorming as a popular form of entertainment.

Despite the decline of barnstorming, its impact on aviation was profound. Many aviators who started their careers as barnstormers went on to achieve significant milestones in the field. Charles Lindbergh, the first pilot to complete a solo nonstop flight across the Atlantic Ocean, was a former barnstormer. Other notable figures who were part of the barnstorming era included daredevil Roscoe Turner, female pilot Bessie Coleman, and Pancho Barnes, a famous speed racer during the "Golden Era of Airplane Racing."

Barnstorming also played a crucial role in shaping the public's perception of aviation. It brought air travel and aerial stunts to rural towns, where people had never seen an airplane up close. The excitement surrounding barnstorming and the remarkable skills of the pilots and performers fostered a greater appreciation and enthusiasm for aviation among the general public. This excitement paved the way for the development of commercial air travel and contributed to the rapid advancements in aviation technology during the 20th century.

Although barnstorming as a widespread phenomenon came to an end in the late 1920s, its influence on aviation history and the subsequent emergence of air shows and aerial displays is undeniable. Today, air shows continue to captivate audiences, with skilled pilots performing breathtaking stunts and showcasing the latest advancements in aviation technology. These modern air shows can trace their roots back to the daring feats and showmanship of the barnstormers who once captivated rural America and inspired a nation to look to the skies.

1.6.2 Modern Air Shows: From Exhilarating Entertainment to Episodic Tragedies

On a fateful Saturday afternoon, July 27, 2002, at the Sknyliv airfield near Lviv, Ukraine, a breathtaking spectacle turned into a heart-wrenching tragedy. More than 10,000 eager spectators gathered to witness the incredible aerobatic feats of a Sukhoi SU-27 fighter jet, skillfully flown by experienced pilots Volodymyr Toponar and Yuriy Yegorov. The event, celebrating the 60th anniversary of the Ukrainian Air Force's 14th Air Corps, would soon become an unforgettable nightmare (Radio Free Europe/ Radio Liberty, 2022; Sydney Morning Herald, 2002).

As the jet performed a daring spiraling descent at low altitude, the audience watched in awe, captivated by the display of precision and speed. Suddenly, the aircraft executed an unplanned roll, its left wing dropping and clipping the treetops. The jet careened toward the ground, skimming the surface before colliding with an Ilyushin Il-76 transport aircraft. In an instant, the fighter jet exploded and cartwheeled into the horrified crowd.

The pilots narrowly escaped death, ejecting just before impact and suffering only minor injuries. Tragically, 78 unsuspecting spectators, including 28 innocent children, were not as fortunate. They perished in the chaos of explosions and debris, while hundreds more were left with head injuries, burns, and broken bones. The incident left a profound impact on the survivors, with approximately 543 people experiencing physical and emotional trauma (Kozyrieva, 2009).

In the aftermath, the search for answers and accountability began. Ukrainian President Leonid Kuchma publicly blamed the military, leading to the dismissal of Air Force head General Viktor Strelnykov (Finlay & Joshi, 2022). The pilots stood trial before a military tribunal, facing charges of gross negligence and attempting unpracticed maneuvers. Despite their claims of receiving an inaccurate flight map and being denied a requested rehearsal, Toponar and Yegorov were found guilty and sentenced to 14 and 8 years in prison, respectively (Frosevych, 2006).

The Sknyliv disaster, often considered the worst air show tragedy in history, serves as a harrowing reminder of the inherent risks and potential human errors involved in performing such high-stakes aerobatics (see Figure 1.17). Yet, the undeniable allure of flight and the thrill of witnessing these gravity-defying feats continue to draw audiences worldwide, captivated by the seemingly impossible brought to life.

1.6.3 Air Show Safety: A Data-Driven Perspective of Psychological Factors That Influence Resilient Safety Culture in Air Shows

Having introduced you to the interesting world of the air show industry from a historical perspective to current times, we now establish the rationale for this book; with a seeming paucity of literature and data-driven research on how psychological

FIGURE 1.17 Ukrainian Air Force Su-27 is performing at RIAT 2011. Copyright: Tim Felce.

risk factors influence air show performers' safety, we conducted a study as part of a doctoral dissertation to assess the relationships between risk perceptions, risk tolerance, hazardous attitudes and mindfulness on-air show performers resilient safety culture.

Over a period of 8 weeks during the 2021 international air show display season, the perspectives of stakeholders in the international air show community (air show performers and air bosses) were sampled using anonymous survey instruments, numerous in-person semi-structured interviews, and focus groups. A comprehensive analysis of existing air show-related regulations, guidelines, and occurrence data were analyzed.

The contents of this book are based on significant findings from this research, and it also provides critical implications for further research, policy, and practices in the global air show industry. Specifically, this book aims to enhance safety and optimize air show performances by providing a deeper understanding of the factors influencing air show safety. It is our fervent hope that the reader will derive value through the implementation of targeted interventions and training programs to mitigate risks and improve safety outcomes in the air show industry.

NOTE

1. Emmanouil Argyropoulos performed with a Nieuport IV.G. the first ever flying display in Greece on February 8th 1912, becoming a legendary form in Greek aviation history (Hellenic Air Force (HAF), n.d.).

LIST OF REFERENCES

Agência Nacional de Aviação Civil (ANAC). (2015). *Relatório de análise das contribuições referentes à audiência pública no 17/2015, do regulamento Brasileiro da aviacao civil no 91 e do egulamento Brasileiro da aviacao civil no 01.* Agência Nacional de Aviação Civil (ANAC). https://www.gov.br/anac/pt-br/acesso-a-informacao/participacao-social/consultas-publicas/audiencias/2015/17/ap172015rac.pdf/view

Agência Nacional de Aviação Civil (ANAC). (2023, April 4). *National Civil Aviation Agency.* Agência Nacional de Aviação Civil (ANAC). https://www.gov.br/anac/en/national-civil-aviation-agency

Air Accidents Investigation Branch. (2017). *Report on the accident to Hawker Hunter T7, G-BXFI near Shoreham airport on 22 August 2015* (Aircraft Accident Report No. 1/2017; p. 452). Department of Transport, United Kingdom. https://assets.publishing.service.gov.uk/media/58b9247740f0b67ec80000fc/AAR_1-2017_G-BXFI.pdf

Aircraft Owners and Pilots Association (AOPA). (2009, July 5). *The "blackbirds": The story of the first all-black aerobatic team.* https://www.aopa.org/news-and-media/all-news/2009/july/pilot/the-blackbirds

Baker, D. (1994). *Flight and flying: A chronology.* Facts on File. http://archive.org/details/flightflyingchro0000bake

Bremner, C. (2008, April 7). *French pilots show women can fly.* Times Online - WBLG. https://web.archive.org/web/20080407080433/http://timescorrespondents.typepad.com/charles_bremner/2008/04/post.html

British Gliding Association. (2023). British Gliding Association. https://www.gliding.co.uk/

British Hang Gliding and Paragliding Association. (2023). British Hang Gliding and Paragliding Association. https://www.bhpa.co.uk/

Casey, L. S. (1981). *Curtiss, the hammondsport era, 1907–1915.* Crown Publishers. http://archive.org/details/curtisshammondsp0000case

Civil Aviation Safety Authority. (2022). *AC 91-21, air displays.* Australian Government. https://www.casa.gov.au/air-displays

Dekker, S. (2017). *The field guide to understanding "human error"* (3rd ed.). CRC Press. https://www.taylorfrancis.com/books/mono/10.1201/9781317031833/field-guide-understanding-human-error-sidney-dekker

European Airshow Council. (2023). *EAC: Flying display directors safety seminar.* EAC: Flying Display Directors Safety Seminar. https://www.europeanairshow.org/copy-of-fast-jet-safety-workshop

Federal Aviation Administration. (2016). *Pilot's handbook of aeronautical knowledge* (2016th ed.). Federal Aviation Administration. https://www.faa.gov/regulations_policies/handbooks_manuals/aviation/phak/media/pilot_handbook.pdf

Fédération Aéronautique Internationale (FAI). (2019, July 28). *Ninety years since Maryse Bastié became the first woman to set an aviation world record.* https://www.fai.org/news/ninety-years-maryse-basti%C3%A9-became-first-woman-set-aviation-world-record

Finlay, M., & Joshi, G. (2022, August 8). *Ramstein and Sknyliv: The world's deadliest air show disasters.* Simpleflying. https://simpleflying.com/ramstein-sknyliv-disasters/

French Civil Aviation Authority. (2023). *French Civil Aviation Authority, Directorate General for Civil Aviation (DGAC).* Ministry of Ecological Transition. https://www.ecologie.gouv.fr/en/french-civil-aviation-authority

Friedlander, M. P., & Gurney, G. (1973). *Higher, faster, and farther.* Morrow. http://archive.org/details/higherfasterfart0000frie

Frosevych, L. (2006, January 28). *Prosecutorial aerobatics after the Skynliv tragedy* [Obozrevatel - Translated from the original article in Ukrainian]. https://www.obozrevatel.com/ukr/author-column/17369-prokurorskij-pilotazh-pislya-sknilivskoi-tragedii.htm

Harris, S. (1991). *The first to fly: Aviation's pioneer days*. Tab/Aero Books. http://archive.org/details/firsttoflyaviati0000harr

Hellenic Air Force (HAF). (n.d.). *First airplane to ever fly in Greece*. Hellenic Air Force. Retrieved May 7, 2023, from https://www.haf.gr/en/history/historical-aircraft/nieuport-ivg/

International Council of Air Shows. (2023a). *Air Boss Recognition Program (ABRP) Manual*. International Council of Air Shows. https://airshows.aero/GetDoc/4140

International Council of Air Shows. (2023b). *ICAS: Air Boss Academy*. ICAS: Air Boss Academy. https://airshows.aero/CMS/AirBossAcademy

Jablonski, E. (1980). *Man with wings: A pictorial history of aviation*. Doubleday. http://archive.org/details/manwithwingspict0000jabl

Janis, I. L. (1972). *Victims of groupthink: A psychological study of foreign-policy decisions and fiascoes* (pp. viii, 277). Houghton Mifflin.

Kozyrieva, T. (2009). *Two Sknyliv boys seven years later*. У К Р А Ї Н С Ь К А П Р Е С-Г Р У П А. https://day.kyiv.ua/en/article/society/two-sknyliv-boys-seven-years-later

Matshili, R. (2019, May 27). *Major Mandisa Mfeka flying high at inauguration*. Pretoria News. https://www.iol.co.za/pretoria-news/news/major-mandisa-mfeka-flying-high-at-inauguration-24180047

McIntyre, G. (2017, August 8). *Anna Serbinenko: From Swiss banker to Sky Dancer*. Vancouversun. https://vancouversun.com/news/local-news/anna-serbinenko-from-swiss-banker-to-sky-dancer

MistyBlues. (2023). *Meet the Mistys*. MistyBlues.Net. https://mistyblues.net/meet-the-mistys/

National Air and Space Museum. (2021, November 1). *First American women in flight*. https://airandspace.si.edu/stories/editorial/first-american-women-flight

Neubeck, K. (2020). *The Thunderbirds: The United States Air Force's flight demonstration team, 1953 to the present*. Schiffer Military. https://www.goodreads.com/book/show/51343685-the-thunderbirds

Nickerson, R. S. (1998). Confirmation bias: A ubiquitous phenomenon in many guises. *Review of General Psychology*, *2*(2), 175–220. https://doi.org/10.1037/1089-2680.2.2.175

Onkst, D. H. (n.d.). *Early exhibition aviators*. U.S. Centennial of Flight Commission. Retrieved May 7, 2023, from https://www.centennialofflight.net/essay/Explorers_Record_Setters_and_Daredevils/early_exhibition/EX7.htm

Oshinuga, A. (2023). *Air Oshinuga*. Air Oshinuga. https://www.anthonyoshinuga.com/

Pujol, C. (2014). Maryse Bastié, pionnière de l'aviation française: Entre exploits sportifs et engagement patriotique. *Le Mouvement Social*, *248*(4), 67–81. https://doi.org/10.3917/lms.248.0067

Prendergast, C. (1980). *The first aviators*. Time-Life Books. http://archive.org/details/firstaviators00pren

Przybylski, A. K., Murayama, K., DeHaan, C. R., & Gladwell, V. (2013). Motivational, emotional, and behavioral correlates of fear of missing out. *Computers in Human Behavior*, *29*, 1841–1848. https://doi.org/10.1016/j.chb.2013.02.014

Radio Free Europe/ Radio Liberty. (2022, July 27). *Remembering Skynliv: The deadliest air show disaster in history*. https://www.rferl.org/a/remembering-skynliv-deadliest-air-show-aviation-history/31961756.html

Reichhardt, T. (2010, September 1). *Pégoud flies upside down*. Air & Space Magazine, Smithsonian Magazine. https://www.smithsonianmag.com/air-space-magazine/pegoud-flies-upside-down-1913-143766023/

RIAT. (2020, March 3). *Pilot profile: Kirsty Murphy—First female Red Arrow.* The Royal International Air Tattoo. https://www.airtattoo.com/news/2019/mar/08/pilot-profile-kirsty-murphy-first-female-red-arrow

Rich, D. L. (1993). *Queen Bess: Daredevil aviator.* Smithsonian Institution Press. http://archive.org/details/queenbessdaredev00rich

Schmader, T., Johns, M., & Forbes, C. (2008). An integrated process model of stereotype threat effects on performance. *Psychological Review, 115*(2), 336–356. https://doi.org/10.1037/0033-295X.115.2.336

South African Civil Aviation Authority. (2021). *The South African special air events handbook.* South African Civil Aviation Authority. https://caasanwebsitestorage.blob.core.windows.net/ga-tgm/The%20South%20African%20Special%20Air%20Events%20Handbook.pdf

Slotnik, D. E. (2019, December 11). Overlooked No More: Bessie Coleman, Pioneering African-American Aviatrix. The New York Times. https://www.nytimes.com/2019/12/11/obituaries/bessie-coleman-overlooked.html

Steele, C. M. (1997). A threat in the air. How stereotypes shape intellectual identity and performance. *The American Psychologist, 52*(6), 613–629. https://doi.org/10.1037//0003-066x.52.6.613

Stone, R. (Director). (2021, November 15). *Blue Angels: Around the world at the speed of sound* [2 hours and 10 minutes]. https://www.amazon.com/Blue-Angels-Around-World-Special/dp/B07RD1PTL2/ref=tmm_aiv_swatch_0?_encoding=UTF8&qid=&sr=

Sydney Morning Herald. (2002, July 28). *78 killed in deadly air-show crash.* https://www.smh.com.au/world/78-killed-in-deadly-air-show-crash-20020728-gdfhq8.html

The Red Baron Team. (2023). *The Red Baron Team.* The Red Baron Team. https://www.red-baron.com.au/our-team

Thompson, S. C. (1999). Illusions of control: How we overestimate our personal influence. *Current Directions in Psychological Science, 8*, 187–190. https://doi.org/10.1111/1467-8721.00044

UAE General Civil Aviation Authority. (2023). *UAE General Civil Aviation Authority.* UAE General Civil Aviation Authority. https://www.gcaa.gov.ae/en/home

UK Air Accidents Investigation Branch. (2017). *Aircraft accident report AAR 1/2017—G-BXFI, 22 August 2015* (Aircraft Accident Report AAR 1/2017). UK Air Accidents Investigation Branch. https://www.gov.uk/aaib-reports/aircraft-accident-report-aar-1-2017-g-bxfi-22-august-2015

UK Civil Aviation Authority. (2023a). *CAP 403: Flying displays and special events: Safety and administrative requirements and guidance.* UK Civil Aviation Authority. https://publicapps.caa.co.uk/docs/33/CAP403%20Edition%2020.pdf

UK Civil Aviation Authority. (2023b). *Civil Aviation Authority.* Civil Aviation Authority. https://www.caa.co.uk/home/

U.S. Air Force. (2023). *U. S. Air Force Demonstration Team: The Thunderbirds.* US Air Force. https://www.airforce.com/thunderbirds/overview

U.S. Navy. (2023). *U.S. Navy Blue Angels.* U.S. Navy Blue Angels. https://www.blueangels.navy.mil/

Veronico, N. A. (2005). *The Blue Angels: A fly-by history, Sixty years of aerial excellence.* Motorbooks International. https://openlibrary.org/books/OL8011569M/The_Blue_Angels_A_Fly-By_History

Wang, L., & Zhang, J. (2020). The effect of psychological risk elements on pilot flight operational performance. *Human Factors and Ergonomics in Manufacturing & Service Industries, 30*(1), 3–13. https://doi.org/10.1002/hfm.20816

Whitehouse, A. (1965). *The early birds: The wonders and heroics of the first decades of flight.* Doubleday & Company, Inc. http://archive.org/details/earlybirdswonder0000arch

You, X., Ji, M., & Han, H. (2013). The effects of risk perception and flight experience on airline pilots' locus of control with regard to safety operation behaviors. *Accident Analysis and Prevention*, *57*, 131–139. https://doi.org/10.1016/j.aap.2013.03.036

FURTHER READING

Books

Brooks-Pazmany, K. (1991). *United States women in aviation, 1919–1929.* Smithsonian Institution Press.
Caidin, M. (1991). *Barnstorming.* New York: Bantam Books.
Cleveland, C. M. (1978). *Upside-down Pangborn: King of the barnstormers.* Aviation Book Company.
Cooper, A. L. (1993). *On the wing: Jessie Woods and the Flying Aces Air Circus.* Black Hawk Publishing Co.
Corley-Smith, P. (1989). *Barnstorming to bush flying: British Columbia's aviation pioneers, 1910–1930.* SONO NIS Press.
Corn, J. J. (1983). *The winged gospel: America's romance with aviation, 1900–1950.* Oxford University Press.
Dwiggins, D. (1966). *The air devils: The story of balloonists, barnstormers, and stunt pilots.* J.B. Lippincott Company.
Freydberg, E. H. (1994). *Bessie Coleman: The Brownskin Lady Bird.* Garland Publishing, Inc.
Gibbs-Smith, C. H. (1970). *Aviation: An historical survey from its origins to the end of World War II.* Her Majesty's Stationery Office.
Glines, C. V. (1995). *Roscoe Turner: Aviation's master showman.* Smithsonian Institution Press.
Hardesty, V., & Pisano, D. (1984). *Black wings: The American Black in aviation.* Smithsonian Institution Press.
Hart, P. S. (1992). *Flying free: America's first Black aviators.* Lerner Publication Company.
Hatfield, D. D. (1976). *Dominguez air meet.* Northrop University Press.
Launius, R. D., & Embry, J. L. (1996). *The 1910 Los Angeles airshow: The beginnings of air awareness in the west.* Southern California Quarterly.
Lerchner, J. V. (Ed.). (2003). *The Allyn & Bacon anthology of traditional literature.* Allyn & Bacon/Longman.
March, J. R. (2014). *Dictionary of classical mythology* (2nd ed.). Oxbow Books.
Marrero, F. (1997). *Lincoln Beachey: The man who owned the sky.* Scottwall Associates.
Moolman, V. (1981). *Women aloft.* Time-Life Books.
O'Neil, P. (1981). *Barnstormers and speed kings.* Time-Life Books.
Reinhardt, R. (1995). *Day of the daredevil.* American Heritage of Invention & Technology.
Rhode, B. (1970). *Bailing wire, chewing gum and guts: The story of the Gates Flying Circus.* Kennikat Press.
Rich, D. L. (1998). *The magnificent Moisants: Champions of early flight.* Smithsonian Institution Press.
Ronnie, A. (1973). *Locklear: The man who walked on wings.* A.S. Barnes and Company.
Tessendorf, K. C. (1988). *Barnstormers and daredevils.* Atheneum.

CHAPTERS IN BOOKS

Jenkins, A. C. (1975). *Barnstormers.* In Airborne. Blackie.

WEBSITES

Barnstorming and air mail. Prairie Public. http://www.prairiepublic.org/features/RRRA/air.html

Barnstorming and early pilots in the mid 27. Columbia area. http://www.angelfire.com/me/mcalch/barn2.html

Barnstorming to bush flying. Stuart Graham Papers. http://collections.ic.gc.ca/sgraham/barn.htm

Bessie Coleman. Allstar Project. http://www.allstar.fiu.edu/aero/coleman.html

Bessie Coleman. Public Broadcasting System. http://www.pbs.org/wgbh/amex/flygirls/peopleevents/panbeAMEX02.html

Bessie Coleman. ThinkQuest Library. http://library.thinkquest.org/10320/coleman.htm?tqskip1=1&tqtime=0313

Biography: Bessie Coleman. ThinkQuest Internet Challenge Library. http://library.thinkquest.org/21229/bio/bcole.html

Charles H. Hubbell, 1899–1971. Barnstormer Art. http://www.barnstormer.com/charleshubbell.htm

Earle, J. (n.d.). *Barnstorming pilots always drew a crowd.* Wings Over Kansas. http://www.wingsoverkansas.com/history/aviation-pioneers/barnstorming.html

Fun facts about barnstormers. http://www.angelfire.com/me/mcalch/barn2.html

Information about and how to fly the Curtiss Jenny barnstormer. Fiddler's Green. http://www.fiddlersgreen.net/aircraft/WWI/jenny/jenn_info/jenn_info.html

International society of aviation barnstorming historians. Crossroads Access Corinth History. http://www2.tsixroads.com/Corinth_MLSANDY/isabh.html

McCullough, D. (n.d.). *Daredevil Lindbergh and his barnstorming days.* http://airsports.fai.org/jun2000/jun200005.html or http://www.pbs.org/wgbh/amex/lindbergh/sfeature/daredevil.html

National Air and Space Museum. (1999). *African American pioneers in aviation teacher guide 1920–Present.* https://airandspace.si.edu/files/pdf/explore-and-learn/teaching-posters/aviation.pdf

Oertly, L. (n.d.). The legacy of Bessie Coleman. Federal Aviation Administration. http://www.faa.gov/avr/news/Bessie.html

Scott, P. (2012, June 1). *Blackbirds and the colored air circus of 1931.* Airfact Journal. https://airfactsjournal.com/2012/06/blackbirds-and-the-colored-air-circus-of-1931/

Walt Pierce. American Barnstormer. http://www.Americanbarnstormer.com/people.html

2 Risk Management

2.1 INTRODUCTION

Air shows have long been a source of entertainment and fascination for audiences worldwide. These events showcase the skill and expertise of pilots, as well as the capabilities of various aircraft. However, there is an inherent safety risk posed by hazards associated with such high-energy aerobatic maneuvers at air shows (Barker, 2003). Examples of hazards associated with air shows include high speed and low altitudes required for some maneuvers, extreme proximity of aircraft in formation flights, and high gravitational forces imposed on the performers during tight turns and steep climbs/dives. Ensuring a zero accident or incident air show industry might be an achievable long-term goal, yet the international air show community should understand that errors will occur that could lead to safety occurrences. Consequently, the pragmatic approach is to focus on mitigating the risk posed by these hazards with effective risk management plans in place to a level that is as low as reasonably practicable (ALARP) (Stolzer & Goglia, 2016).

This chapter seeks to identify and assess the risks faced by air show performers, using real-world examples to highlight potential dangers. Additionally, we discuss the strategies and techniques employed by air show performers to manage these risks and enhance overall safety. A key focus will be on the suggestion of an *Airshow Operational Risk Management (AORM)* and an *Airshow-Safety Management System (ASMS)* specifically designed for air show performers. These safety tools aim to systematically identify, assess, and manage hazards while fostering a proactive safety culture. We also address the insidious risk posed by the normalization of deviance, the incremental acceptance and assimilation of atypical conduct or practices deemed unsuitable or perilous (Vaughan, 2016), which can be a precursor of catastrophic events if not properly managed. This chapter explores how *AORM* and *ASMS* initiatives can emphasize strict adherence to safety protocols, facilitate open communication among stakeholders, and promote ongoing learning to prevent complacency and the gradual acceptance of at-risk behaviors such as flying while fatigued or stressed out. By examining these measures, we hope to provide insights into effectively mitigating both apparent and insidious risks, ultimately contributing to increased safety and a reduced likelihood of safety events during air show performances.

2.1.1 WHAT IS A RISK?

Risk is defined as the probability or likelihood of an adverse event or outcome occurring, along with the potential consequences or impact associated with that event. It involves the uncertainty of future events and is generally characterized by the

DOI: 10.1201/9781003431879-2

potential for loss, damage, or harm. Practical risk management aims to identify, assess, and prioritize risks to minimize, monitor, and control their potential impact on an individual, organization, or system (Lam, 2014; Tchankova, 2002).

According to the International Civil Aviation Organization's (ICAO) Safety Management Manual (SMM), 2018 edition, risk is defined as: "The probability of an event occurring, combined with the severity of the potential consequences of that event" (ICAO, 2018, p. 4). Furthermore, ICAO characterizes risk management as the process of pinpointing, evaluating, and ranking risks and then efficiently allocating resources to lessen, oversee, and regulate the likelihood and/or consequences of adverse events or to optimize the potential of opportunities (ICAO, 2018, pp. 2–13).

2.1.1.1 Calculated Risk

Taking risks is essential to life, but not all risks are equal. Some risks can lead to significant rewards, while others can result in serious negative consequences. That's where calculated risks come in, requiring an individual to weigh the potential benefits against the potential risks of a particular action.

Calculated risks are not made randomly or recklessly (Kaplan & Mikes, 2012). Instead, they are thoughtful and intentional decisions made after considering the potential outcomes carefully. These decisions are based on the best available information and an individual's knowledge, experience, and judgment. Therefore, calculated risks are a crucial part of aviation decision-making across many fields.

In air shows, calculated risks are critical to ensuring safe events. Air show performers must consider numerous factors before making a decision, such as weather conditions, aircraft performance, and air traffic control. Therefore, air show performers need to be trained to weigh the potential benefits against the potential risks of every decision they make, from choosing a maneuver to selecting the appropriate minimum altitude. By carefully considering the potential outcomes of their decisions, pilots can minimize risks and ensure safe flights.

It's important to note that calculated risks are not always successful, and there is always the potential for adverse outcomes. However, by taking calculated risks, individuals can increase their chances of success and achieve their goals. Additionally, taking calculated risks can help individuals build confidence, improve decision-making skills, and learn from their experiences (Heath & Heath, 2013).

2.1.2 Risk Management among Air Show Performers: An Overview

Air show performances inherently involve risks, making it crucial to employ solid risk management strategies as a fundamental aspect of these events. By conducting appropriate risk management, performers can successfully minimize the risks linked to their aerial routines, safeguarding both themselves and the audience.

2.1.2.1 Risk Identification

Effective risk management in air shows involves identifying potential hazards, assessing their probability and potential impact, and implementing strategies to minimize their occurrence. Air show performers must address various risk factors, such as weather conditions, mechanical failures, human errors, and crowd management, and then apply appropriate risk mitigation strategies.

2.1.2.2 Weather Conditions

Unfavorable weather conditions, such as thunderstorms characterized by gusting winds and poor visibility, can significantly increase the risk of accidents during air shows. Not only do performers need to constantly keep an eye on weather forecasts and modify their routines accordingly to maintain safety, but air show organizers also have a responsibility to monitor weather conditions, relay vital information to the performers, and make necessary adjustments to the event schedule as required. Numerous air shows have made modifications to their schedules, with the 2022 Sanicole Airshow in Belgium serving as a prime example. This event faced alterations and cancellations to flying displays as a result of adverse weather conditions, including low visibility due to rain, as well as gusting winds (Karachalios M., personal communication, September 21, 2022).

Gusts are sudden, brief increases in wind speed that can make it difficult for pilots to maintain control of their aircraft, especially during aerobatic maneuvers. The combination of rain and gusting winds can create a hazardous environment for both pilots and spectators, which is why organizers had to make changes to the event's schedule to ensure safety.

The organizers of the 2022 Sanicole Airshow in Belgium employed a combination of strategies to adapt to adverse weather conditions and ensure a successful event. Some actions that have been taken include delaying the show when the weather was expected to improve later in the day, postponing specific flying displays until conditions were more favorable, or adjusting the start time of the event.

Organizers have also worked with pilots to modify their routines, allowing for only specific maneuvers that could be performed safely in the prevailing weather conditions, such as flat shows only. This involved simplifying routines, lowering the altitude of certain maneuvers, or eliminating high-risk stunts.

Enhancing communication between the air traffic control, pilots, and event organizers had been crucial to ensuring the safe execution of the air show. Additionally, organizers and meteorologists closely monitored weather forecasts and real-time updates to make informed decisions about the show's schedule and the safety of flying displays.

By employing these strategies, the flying control committee and organizers of the 2022 Sanicole Airshow were able to demonstrate resilience, ensuring a successful event despite the challenging weather conditions.

2.1.2.3 Mechanical Failures

Mechanical issues are a critical risk factor during air shows. Performers must conduct thorough preflight inspections and regular maintenance to minimize the possibility of mechanical failures. In response to the 2019 Snowbirds accident caused by a fuel system failure, the team adopted more stringent inspection protocols and increased maintenance frequency to ensure aircraft reliability (Royal Canadian Air Force, 2019).

2.1.2.4 Human Errors

Human errors can have devastating consequences during air shows. To minimize such risks, pilots undergo extensive training and adhere to strict safety protocols. The tragic Blue Angels crash in 2016, which was blamed on pilot error due to deviations in standard operating procedures (SOPs) while executing a Split S maneuver (U.S. Navy, 2016), led to changes in the team's risk management procedures.

2.1.2.5 Crowd Management

Effective crowd management is essential for ensuring spectator safety during air shows. Organizers must establish designated viewing areas, implement crowd control measures, and provide emergency services. In 2011, the Reno Air Races (National Transport Safety Board (NTSB), 2011) experienced a tragic accident in which an aircraft crashed into the spectator area, resulting in numerous fatalities and injuries. Following the incident, the event organizers implemented additional safety measures, including reinforced barriers and increased distance between spectators and the aircraft.

2.1.2.6 Risk Mitigation

Air show performers employ various risk mitigation strategies to ensure the safety of both performers and spectators. These strategies include adhering to established safety standards, utilizing advanced technology, implementing thorough training programs, and enhancing communication.

Especially during the performance, clear and effective communication among pilots, ground crews, and air traffic controllers is essential for maintaining safety during air shows. For instance, the Thunderbirds, the U.S. Air Force's aerial demonstration team, employ a standard practice among military teams for a combination of radio communications and hand signals to ensure precise coordination during their performances.

2.2 RISKS IDENTIFIED IN LITERATURE AND ACCIDENT REPORTS (*MOST FREQUENTLY OBSERVED RISKS*)

Numerous risks for air show performers have been identified in the literature and accident investigation reports. Using the 3M accident taxonomy in Barker's (2020a) analysis of air show accidents, a summary table of identified risks has been developed and can be seen in Table 2.1.

However, the most frequently observed risks during air shows that have contributed to accidents or incidents throughout the years can be concluded in the following:

TABLE 2.1
Risks Identified in Literature and Accident Reports, Categorized by Factor (Barker, 2020a)

	huMan	Machine	Medium
1	Loss of Situational Awareness	Wing Failure	Bird Strike
2	Fatigue	Flutter	Wind
3	Human Error	Tailplane	Wake Vortex
4	Incapacitation	Engine Failure	Cloud
5	Distraction	Flight Controls	Foreign Object Debris (FOD)
6	Mid-Air Collision	Under Carriage Door	Turbulence
7	Loss Of Control	Canopy Unlocked	Runway Condition
8	–	Parachute Failure	Downdraughts
9	–	Wing-walking Rope Failure	Passenger Interference
10	–	–	Bomb Collision
11	–	–	Misting in the Environment Control System
12	–	–	Pyrotechnics

2.2.1 Loss of Control In-flight (LOC-I)

Loss of control in-flight (LOC-I) is a leading cause of accidents at air shows and can result from various factors, including pilot error, mechanical failure, or adverse weather conditions. One such incident occurred at the 2011 Reno Air Races when a P-51 Mustang, named "The Galloping Ghost," experienced a catastrophic structural failure when the aircraft's elevator trim tab separated during a high-speed, low-altitude turn. The separation of the trim tab led to a rapid pitch-up, causing the aircraft to experience extreme G forces that resulted in the in-flight breakup of the airplane and subsequent uncontrolled descent, resulting in 11 fatalities and 74 injuries (National Transport Safety Board (NTSB), 2011).

Another notable example occurred in 2022 when one of the last Hawker Hurricanes flying crashed during an air show in the Czech Republic after a hammerhead turn[1] was initiated with low energy, resulting in a subsequent stall during the recovery and, due to insufficient altitude, the aircraft impacted the ground leading to the pilot's death. Fortunately, no spectators were injured, but the event emphasized the dangers associated with LOC-I at air shows (Aviation Safety Network, 2022).

2.2.2 Vertical Maneuvers

Vertical maneuvers, such as loops and climbs, can pose significant risks to pilots during air shows. These high-energy maneuvers can result in high G forces, which can cause pilots to experience disorientation, spatial disorientation, or even loss of consciousness (G-LOC) (Newman, 1997). To mitigate this risk, pilots undergo intensive training in high G environments and may use specialized equipment, such as G suits, to maintain blood flow and consciousness during high G maneuvers (Federal Aviation Administration, 2022).

2.2.3 Energy-Depleting Maneuvers

Energy-depleting maneuvers, such as tight turns and sudden changes in altitude, can rapidly consume an aircraft's kinetic and potential energy reserves, leading to a loss of control or even a stall (Federal Aviation Administration, 2021). In 2015, a Hawker Hunter aircraft crashed during the Shoreham Airshow in the UK after the pilot failed to complete a loop-the-loop maneuver, possibly due to insufficient energy (UK Air Accidents Investigation Branch, 2017). Pilots must carefully manage their aircraft's energy, both kinetic and potential, during air show performances and ensure adequate energy reserves before attempting such maneuvers.

2.2.4 Flight into Terrain (FIT) or Flight into Object (FIO)

Air show accidents involving collisions with terrain or objects can arise from low-altitude maneuvers, misjudging distances, or flying too close to spectators (Wiegmann & Shapell, 2016). A tragic example occurred during the 2009 Air Show at Sknyliv in Ukraine when a Ukrainian Air Force Su-27 Flanker fighter jet first hit a tree, and then some aircraft part of the static display, ending up in the spectators' area, killing 83 people and injuring 545 others (BBC News, 2002; Radio Free Europe/ Radio Liberty, 2022). Another incident transpired in 2022 at an Air Show in Argentina when a DHC-1 Chipmunk hit a pole inside the crowd line after flying too low. This tragic event resulted in four injuries (Aero-News Network, 2022).

2.2.5 Mid-Air Collision (MAC)

Mid-air collisions during air shows typically involve formation flying or aerobatic displays, posing significant risks. In 2007, at the Radom Air Show in Poland, two planes from the Zelazny Aerobatics Team collided mid-air, killing both pilots (Aviation Safety Network, 2007). Another mid-air collision happened in 2022 between a B-17 and a P-63 of the Commemorative Air Force (CAF), killing all six aircrews onboard (National Transport Safety Board, 2022).

2.2.6 Human Factors (HF)

In both the 1994 B-52 accident and the 2005 Pilatus PC-21 crash, human factors were significant contributors to the accidents. Hazardous attitudes, pilot fatigue, and chronic stress can negatively impact pilot performance during air shows, increasing the likelihood of accidents (Karachalios, 2022).

The U.S. Airforce B-52 accident in 1994 was a tragic event that occurred during the Fairchild Air Force Base Open House and Airshow in Washington State, USA. The aircraft crashed while performing a low-altitude maneuver during a practice flight for the air show (USAF, 1994). All four crew members on board the aircraft were killed in the crash. The subsequent investigation conducted by the U.S. Air Force revealed several contributing factors to the accident. The primary cause was identified as the pilot's failure to maintain adequate airspeed, leading to a stall and loss of control during the maneuver (U.S. Air Force, 1994). Additionally, human factors played a significant role in the accident, including the pilot's overconfidence and aggressive flying style, and inadequate supervision by Air Force leadership.

On 13 January 2005, one of the two Pilatus PC-21 development aircraft with registration HB-HZB crashed in Buochs, Switzerland, while conducting an aerobatic training flight. In this Pilatus PC-21 crash, the accident was attributed to pilot error, according to an official accident investigation report from the Swiss aviation accident investigation bureau BFU (2005). The aircraft, while practicing a formation display as number two of a pair of PC-21s, made contact with the ground with its left wing, causing it to ignite and become engulfed in flames. The pilot tragically lost his life, and the plane was entirely destroyed. The report identified several factors that may have contributed to the accident. These include the impairment of the pilot's vision, the time pressure and multiple tasks imposed on the pilot, and the difficulty of the maneuver being flown at a low level. These factors can be linked to human factors such as stress, fatigue, and distraction, which can greatly affect a pilot's ability to perform complex tasks.

2.2.7 Push-Pull Effect

The push-pull effect occurs when a pilot experiences rapid changes in G forces, transitioning from positive G forces (pushing the pilot into their seat) to negative G forces (pulling the pilot out of their seat) (Metzler, 2020). These rapid changes can cause physiological stress and discomfort, increasing the risk of pilot error. To mitigate the push-pull effect, pilots must include such effects in the planning of their display sequence. Then they must maintain smooth and controlled inputs on the aircraft's controls and be familiar with the physiological sensations associated with rapid changes in G forces. The accident involving Thunderbird number 4 of the U.S. Air Force Air Demonstration Squadron on April 4, 2018, illustrates this phenomenon's devastating outcomes (U.S. Air Force, 2018). The pilot was doing a Split S maneuver when he suffered a G-induced loss of consciousness and total incapacitation. The aircraft crashed, fatally injuring the pilot without an ejection attempt. The Accident Investigation Board determined that the cause of the mishap was the pilot's G-LOC during the Split S portion of the maneuver. Two factors substantially contributed to the mishap: the pilot's diminished tolerance to +G's due to the push-pull effect and a decrease in the effectiveness of the pilot's Anti G straining maneuver under those conditions.

2.2.8 Distractions in the Cockpit

Distractions in the cockpit during air shows can increase the risk of accidents. These distractions may include unexpected system alerts, radio communication issues, or even insects entering the cockpit. To minimize cockpit distractions, pilots must maintain situational awareness and have a clear understanding of their aircraft's systems and procedures, allowing them to respond effectively to any unexpected occurrences.

2.2.9 Mechanical Failure

Mechanical failures, such as engine malfunctions or structural damage, can lead to accidents at air shows. In 2006, a vintage Hawker Hunter experienced mechanical failure and crashed during the Oregon International Airshow Hillsboro, resulting in the death of the pilot (Casey, 2007). The 2013 forced landing of an MXS Aircraft at the Quad City Airshow in Davenport was also attributed to engine failure (Aero-News Network, 2013).

2.2.10 Bird Strikes

Bird strikes are a known risk in aviation and can also pose a threat during air shows. Ingesting a bird into an aircraft's engine or striking a bird during high-speed maneuvers can cause significant damage, potentially leading to loss of control (Dolbeer, 2011). In 2022, during a practice session, a bird strike incident with an NF-5 aircraft flown by the Turkish Stars aerobatic team of the Turkish Air Force resulted in both engines failing and the pilot safely ejecting from the aircraft (Daily Sabah, 2022).

Air show organizers can reduce the risk of bird strikes by coordinating with local wildlife authorities to deter birds from the performance area and adjusting flight paths to avoid known bird habitats.

2.3 RISK MANAGEMENT PLANNING FOR AIR SHOW PERFORMERS

Air show performers, encompassing both civilian and military participants, confront substantial risks during aerial exhibitions. To guarantee their safety, it is crucial for performers to establish and follow comprehensive risk management plans. An integral aspect of risk management is understanding and balancing risk perception and risk tolerance among air show performers. In this section, we present and discuss risk management strategies implemented by air show performers, with particular emphasis on their risk perception, risk tolerance, and the various measures they employ to minimize potential hazards and enhance safety during performances.

2.3.1 Risk Perception

> According to Sjoberg, Moen, and Rundmo (2004, p. 8), "risk perception is the subjective assessment of the probability of a specified type of accident happening and how concerned we are with the consequences." Hunter (2002) had earlier described risk perception as the awareness of the risk inherent in a situation, implying that risk perception could be influenced by the type of situation and the characteristics of the pilot involved. Martinussen and Hunter (2018) considered risk perception a cognitive activity that accurately assesses internal and external states.

In a study assessing pilots' risk perceptions and decision-making relating to adverse weather scenarios, O'Hare (1990) suggested that an unreasonable estimation of the risks involved could be a factor in pilots' decision to press on into deteriorating weather. The relationship between risk perceptions and risk acceptance has been suggested in a study where a significant negative correlation in terms of risk perception and risk acceptance was observed among general aviation pilots in Australia who elected to fly a risky flight scenario (Drinkwater & Molesworth, 2010). Joseph and Reddy (2013), in a study conducted among Indian army helicopter pilots, found that lower risk perceptions were associated with higher risk-taking tendencies and higher risk attitude scores.

In another risk perception study of Chinese airline pilots, You, Ji, and Han (2013) found that high levels of risk perception and an internal locus of control increase the likelihood of a pilot engaging in safety-oriented activities, including enhanced situation awareness and efficient decision-making. All in all, inadequate risk assessment can lead to poor decision-making, ending in fatal aviation accidents (Efthymiou et al., 2021; Jensen & Benel, 1977).

2.3.2 Risk Tolerance

> Hunter (2002) suggested that risk perception and risk tolerance are two related but often confused constructs. Risk tolerance, according to Hunter, is "the amount of risk that an individual is willing to accept in the pursuit of some goal" (2002, p. 3). Risk tolerance is influenced by a person's general risk aversion as well as the personal value attached to the goal of a given situation (Martinussen & Hunter, 2018).

The concept of risk tolerance exists in aviation and the finance industry. Numerous studies on financial risk tolerance highlight the relationship between risk and reward individuals may seek. In their studies, Gibson et al. (2013) and Hallahan et al. (2004) suggest that age has an inverse relationship with risk tolerance. Hallahan et al. further suggest that marital status, number of dependents, income, and total wealth had a significant relationship with an individual's risk tolerance scores.

So, what is the risk tolerance of air show performers? While it's difficult to quantify this in exact terms, it's clear that air show performers have a higher risk tolerance than the average person, given the nature of their profession. However, this does not mean that they are reckless or cavalier about safety. Rather, it means that they have a unique set of skills, training, and experience that allows them to manage risk effectively and perform at the highest levels of their profession.

Air show performers undergo rigorous training that covers not only the technical aspects of flying but also safety procedures, emergency protocols, and risk management strategies. They also work closely with ground crews, air traffic controllers, and other support personnel to ensure their performances are executed safely and according to plan. Such an interaction demonstrates their commitment to safety and willingness to accept the risks associated with their profession while minimizing them as much as possible.

Despite the likelihood of accidents and other safety events, air show performers periodically accept high risk compared to other aviation-related professionals. This is due to the complexity and challenges associated with intricate aerobatic maneuvers, time constraints, and demanding routines. The acceptance of such high operational risk is normally commensurate with stringent standard operating procedures, detailed safety plans, and briefs, and rigorous training regimens seeped into a culture of excellence.

2.3.3 Psychological Risk Factors Associated with the Air Show Operations

As discussed earlier, the risk associated with psychological factors such as hazardous attitudes, risk perceptions, tolerance, and mindfulness strategies plays a crucial role in shaping the decision-making process of air show performers, which impacts their behaviors both prior to and during their performances. We sought, as part of our research objectives, to unearth and understand these risks deemed significant by performers throughout the global air show industry.

It is worth noting that risk perception can differ considerably among performers and is subject to various influencing factors. In order to better understand these differences, we also sought to identify the factors contributing to these variations in risk perception, with a special focus on those related to demographic characteristics.

2.3.3.1 Qualitative Assessment of Air Show Performers' Perspectives

An extensive qualitative research approach using numerous interviews and focus group sessions of air show performers from various parts of the globe was made. The collected data in the form of audio recordings and reflexive notes were transcribed, validated, and analyzed using a deductive coding/themeing approach. Air show performers' perspectives on four primary areas: existing hazardous attitudes, risk perception and tolerance, mindfulness strategies, and the overall perception of resilient safety culture in the air show community were assessed, and interpretations of the findings were explored.

The first set of questions in the interviews focused on risk perception and tolerance. The questions assessed interviewees' general perceptions of the most significant risks that adversely impact air show displays and the types of risks they were willing to accept when flying in an air show. The themes under the risk perception and risk tolerance areas of study were financial risk, the level of air show display flying risk, an unexpected situation, zero tolerance, operational risk management, and the 5Ms, i.e., the human, the machine, the medium, the management, and the mission. Figures 2.1 and 2.2 show theme maps for the risk perception and risk tolerance areas of focus, respectively.

2.3.3.1.1 Financial Risk

Financial risk has been identified as a hidden risk that impacts all aspects of aerial event management and affects both civilian and military air show performers. Financial constraints primarily influence aircraft maintenance, training, and pilot currency. Insufficient funds can affect all operational levels and have been recognized as a contributing factor in aviation accidents and incidents.

We found out from some interviewees that limited financial resources and competition for these resources pose a potential hazard to air show safety, potentially causing operational strain on performers and organizers and affecting the hiring of safety observers. Additionally, financial resources often compete between operational logistics, such as fuel and lubricants, and safety controls, like collision avoidance technology.

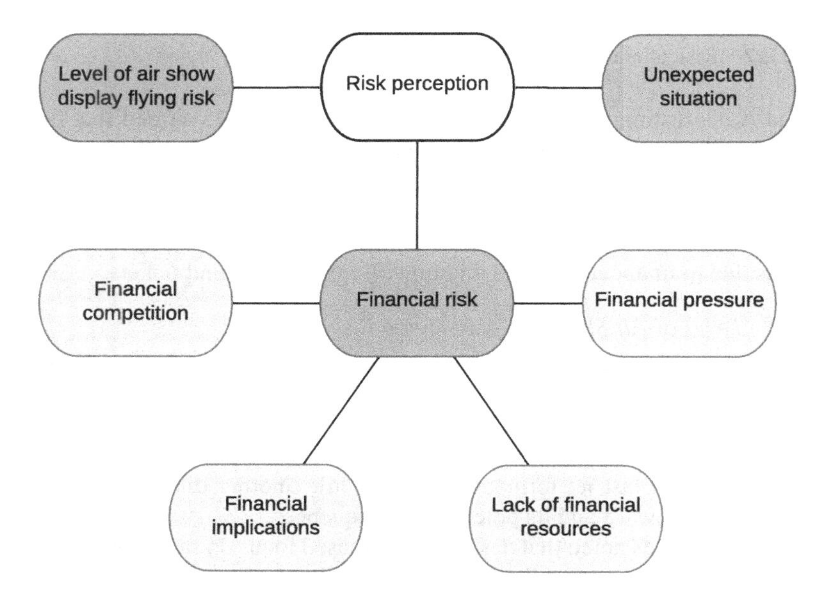

FIGURE 2.1 Risk perception theme map.

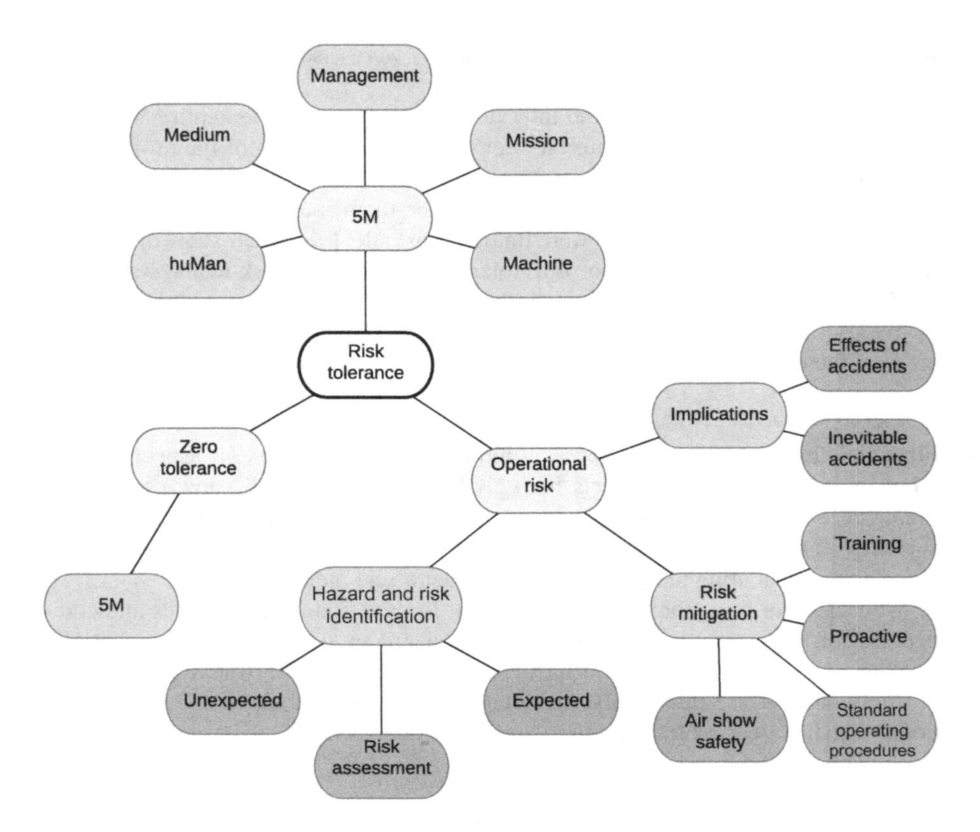

FIGURE 2.2 Risk tolerance theme map.

It is also interesting to note that most air show performers agreed that financial pressures could be experienced at the personal level of air show performers, with one quote emphasizing that insufficient resources for proper training can be a risk. An interviewee noted that financial risk could be associated with air show events themselves, sharing an example of an organizer who allowed flying in unfavorable conditions due to financial troubles and unwillingness to refund tickets.

2.3.3.1.2 Level of Air Show Display Flying Risk

Most air show performers recognized the heightened risk associated with low-altitude air show flying. Numerous factors can impact performers' performance, and a wide range of mishaps can occur. Unforeseen issues may also arise, requiring high levels of skill, experience, and training to address. An interviewee highlighted the considerable risk in fast jet formation flying, while another discussed the balance between risk and reward and its potential consequences.

Some interviewees noted that despite an increased focus on safety in recent years, fatal accidents still occur, and lives are lost annually. An interviewee with extensive air show experience suggested that the high level of risk involved in air show flying demands appropriate professionalism and risk management from all management levels, from aviation authorities to air show organizers, air bosses, and the performers themselves.

2.3.3.1.3 *Unexpected Situation*

Another significant theme that emerged from the study was dealing with "unexpected situations." Numerous problems may arise during an air show performance that requires exceptional skill, knowledge, professionalism, and sometimes even luck to manage effectively. One of the salient challenges during an air show is dealing with an unexpected traffic intrusion. Unexpected traffic in air show airspace could pose a substantial hazard and result in a high risk of mid-air collision. There were differing views among performers regarding such unexpected traffic in air show airspace. A fast military jet demonstration team member expressed little concern, while a civilian performer flying a propeller aircraft expressed worry about unexpected traffic at less organized air shows. The consensus among performers was that adequate planning, including restricted operating airspace, is crucial to mitigating this risk.

The varying responses from the performers may be a function of the level of air show profile complexity they deal with. Military demonstration teams usually expect high levels of Air Traffic Control (ATC) support, separation, and allocation of an exclusive swathe of airspace to operate (sanitization), which minimizes the likelihood of errant intrusion by non-participating aircraft. Also, the military teams normally participate in larger, well-organized exhibitions.

Solo propeller aircraft performers, on the other hand, may operate in less structured events with limited airspace sanitization capabilities. It was also intriguing what an interviewee reported regarding luck. The quote from this interviewee highlights the point: "Everybody needs a little bit of luck. Let us call it luck, but you cannot control your whole life sometimes." In the air show community, luck is still regarded as a survival factor; however, air bosses and air show performers should leave nothing to fortune, as another interviewee highlighted.

Proactive safety risk anticipation through the use of a flight risk matrix and safety briefing before air shows can help identify latent hazards and associated risks. Periodic stakeholders brainstorming sessions could be essential for developing effective controls to mitigate risk associated with unexpected scenarios to tolerable levels in the industry. Sharing lessons learned from such unexpected safety events through regular "hangar talks," newsletters, and social media among the air show community can create an awareness of the likelihood of such unexpected events and a need for higher levels of vigilance.

2.3.3.1.4 *Zero Tolerance*

Zero tolerance as a theme represents a critical juncture in terms of risk acceptability (red line) that the air show community is unwilling to cross when it comes to operational performances. The zero-tolerance theme has underlying codes framed from the *5M model* of Systems Safety in aviation and aerospace organizations (Harris, 2011), where both engineering and management principles, criteria, and techniques are used to attain acceptable safety risk and in which the system is treated holistically, accounting for interactions among its constituent parts.

Most interviewees agreed that they were willing to accept any risk within their control, emphasizing their professionalism in mitigating risks.

A zero-tolerance approach requires commitment from air show leadership (management) to set out realistic expectations and inspire the air show community to strive for excellence and safety during air shows (mission) by mentoring newcomers, fairly applying rules and regulations, setting strong examples, encouraging open communication, recognizing and rewarding outstanding performance, organizing safety seminars and workshops, and building a sense of community.

These approaches create an environment of continuous learning and improvement, foster a culture of safety and professionalism, and encourage the sharing of ideas and best practices, ultimately contributing to the success and reputation of air shows. In this context, air bosses play a crucial role in formulating pragmatic policies and ensuring accountability for safety among participants (humans). Performers (humans) who intentionally disregard air show SOPs and safety policies risk losing their jobs and may not be rehired as their reputation for being unsafe spreads throughout the close-knit air show community.

Examples of such intentional disregard for safety within the air show community are pilots who, while fatigued or under the influence of alcohol and other controlled substances and for which there is a zero-tolerance approach. Disregarding safety and operational policies relating to maintenance and airworthiness of aircraft used for air shows (machine) could result in mechanical failures leading to incidents or, worse case, accidents during air shows.

Dramatic safety occurrences could have a deleterious impact on operational, financial, and moral status throughout the international air show industry. The zero-tolerance approach is an essential aspect of risk management in the air show community that emphasizes a systemic approach to risk acceptability through:

- Design realistic air show profiles within operational environments that minimize risk to an acceptable level.
- Effective leadership of management to ensure a balance between safety and operational goals.
- Air show performer professionalism that recognizes when not to operate outside prescribed safe envelopes during air shows.
- Human-centric aircraft designs that are strictly maintained per regulatory directives.
- Strict adherence to safety policies and zero tolerance for intentional disregard for safety.
- Continuous monitoring and improvements in operational safety through industry-wide learning from safety events.

2.3.3.1.5 Operational Risk Management

Operational risk management emerged as a theme and involves hazard and risk identification, implications, and mitigation strategies during air show preparation, planning, and execution (see Figure 2.2). Risks were classified into expected and unexpected categories, with expected risks including bird strikes, engine failure, and changes in flying routines. Unexpected risks involved human factors, such as distractions during the flying display and unintended effects of event-induced anxiety leading to errors, structural damage from latent design flaws in aircraft (especially experimental and self-built aerobatic aircraft), unrecoverable loss of control, and crowd control after an accident.

2.3.3.1.6 *Expected Hazards and Associated Risks*

The most common expected hazards and associated risks reported were engine failure and bird strikes, highlighting the importance of air show pilots being aware and proactively having contingency plans to manage any of these during their performances. Such contingency planning enables pilots to be mentally and technically prepared to react quickly and effectively during operational mishaps or abnormal situations without compromising the safety of spectators and themselves.

2.3.3.1.7 *Unexpected Hazards and Associated Risks*

During the interviews, some air show performers expressed concerns about unexpected hazards and their associated risks, particularly those related to the human element. An experienced air show performer underscored the role of hazards associated with the human element: "One has to accept there is a fair amount of risk anyway because it is the human element that always introduces the unexpected into it."

Another unanticipated risk involves crowd control during accidents. An extremely experienced air show performer and air boss mentioned that crowd behavior may not always be predictable, with some people moving toward an accident scene to capture it on camera or offer help. This can hinder crash and rescue personnel in their response efforts.

While it is challenging to predict behavior and reactions during and after an air show accident, air bosses and organizers should proactively consider these risks and develop emergency management strategies such as the preplanning and coordination with an emergency operations center (EOC) consisting of mutual aid partners such as law enforcement agencies, paramedics, crash rescue and fire departments, and airport authorities.

2.3.3.1.8 *Risk Assessment Matrix*

To address the challenges of both expected and unexpected hazards and their associated risks, air show performers suggested the use of a risk assessment matrix to identify hazards and assess risks before an air show. A challenge with this approach is that risk assessment processes differ between military and civilian air show performers. In most military organizations, the decision to participate in an air show involves a higher chain of authority, while in civilian organizations, the decision often rests with the individual performer or their team's policies.

It was interesting that an air show performer expressed concern about the effectiveness of written risk assessments for European air shows: "In Europe nowadays, we have to make a written risk assessment that I think it is not smart; anyone can fill it out for you, just to fill it out. It is not right." This highlights the need for more practical and effective risk assessment processes in the air show community beyond merely fulfilling requirements.

Training in using risk assessment matrices could help performers understand their value and not see them as just another paperwork requirement. Risk assessments should reflect each performer's unique appreciation of operational risk and help recognize unacceptable risks during air show activities. A filled-in ICAS (2012) risk assessment matrix is depicted in Figure 2.3.

2.3.3.1.9 Risk Mitigation

Air show operators discussed mitigating expected and unexpected hazards in air shows by taking a proactive approach during display profile design, incorporating safety buffers, adding altitude pads in vertical aerobatic maneuvers, and initially practicing formation aerobatics with wider separations for new performers. They also suggested employing standard operating procedures with explicit go no go criteria, contracts, what-ifs, and contingency planning to anticipate the unexpected.

One example of integrated contingency planning has a search and rescue helicopter at the air show site or within a 10-mile radius during both practice and the actual air show. Flight training, including programs like the annual upset prevention recovery training (UPRT), was emphasized as essential for instilling risk assessment skills and quick reaction abilities in the event of unexpected risks during a display.

2.3.3.1.10 Importance of the Air Show Safety Briefing

Air show performers also emphasized the importance of air show safety briefings as a risk mitigation measure. These briefings can help identify hazards that cannot be addressed beforehand and ensure that all participants, including emergency responders, are prepared for potential risks. The air show safety briefing is also crucial for setting the tone for performers, crash/rescue/firefighting teams, and first aid/medical teams, enabling them to react effectively and quickly in case of an incident or accident.

2.3.3.1.11 Proactive and Self-Engaging Risk Mitigation

Our study findings indicate that risk assessment and mitigation in air shows are ongoing, dynamic processes that continue during the actual flight in the challenging low-level aerobatic environment. Air show performers must engage in real-time risk assessment and decision-making to ensure safety and efficiency. For example, one air show performer mentioned testing the clouds before flying, especially if they are the first to perform: "If I get in that situation, I will test the clouds myself before I fly, especially if I am first."

Proactive safety practices should be encouraged during daily display routines at air shows. While air show pilots may have safety margins and buffers exceeding required standards, they should only accept risks that are tolerable, given the profile's scope and complexity. This allows them to respond skillfully, accurately, and promptly to both expected and unexpected threats during aerobatics.

ICAS Individual Risk Management/Assessment Worksheet

Show Location: [redacted] Date: 5 Sept 2014

- Enter a "1" in the Tally column for each applicable item
- Add Tally scores to obtain final IRM score
- Compare final score to the IRM index to determine assessed risk for the day's scheduled performance

SCHEDULING ASSESSMENT

CREWMEMBER FACTORS	Tally
"I'M SAFE"	
Illness	
Medication	
Stress	
Alcohol	
Fatigue	
Environment	
Scheduling Total:	0

PRE-PERFORMANCE ASSESSMENT

ENVIRONMENTAL FACTORS	Tally
Cold Weather Ops (< 0 C/ 32 F)	
Hot Weather Ops (>35 C/ 95 F)	
Density Altitude	
Field Elevation	
Crosswinds > 10 kts	1
Ceilings < 3000 ft	
Visibility < 5 SM	
Moderate Turbulence Forecast	
Overwater	
Thunderstorms Forecast	
Bird / AHAS Activity ≥ Moderate	
EXECUTION FACTORS	
Scheduled number of performances per day	1
Night Show	
Aircraft/Vehicle Maintenance Issues	1
Participate in Media Flights	
Pre-Performance Total:	3

Scheduling + Pre-Performance = Final IRM Score

INDIVIDUAL RISK MANAGEMENT INDEX		
0 - 5	Low	Total: 3 (low)
6 - 10	Moderate	
11 - 15	High	
16+	Extreme	

RISK ASSESSMENT CRITERIA

How Do You Feel?	> 90% 0	80-90% 2	< 80% 4
Quality of Crew Rest	GOOD 0	FAIR 4	POOR 8
Consecutive Airshows	≤ Three 0	Four 2	≥ Five 4
Days Since Last Show	< 15 0	15-30 2	> 30 4
Work Stress Level	Low 0	Mod 2	High 4
Months Since Last Practice or Routine	No / >3 4	1-3 2	< 1 0
Months Until Expecting Child	No 0	2-9 2	< 2 4
Relationship Issues	No 0	Minor 4	Major 8
Financial Issues	No 0	Minor 4	Major 8
Additional Distractions	No 0	Minor 4	Major 8

INDIVIDUAL RISK ASSESSMENT

1	
2	
3	
4	
5	
6	
7	
8	
9	
10	

CREW INSTRUCTIONS

- Use risk assessment criteria above to assess personal risk
- Record individual score in corresponding box to the left
- Compare individual scores to personal risk index below
- Record number of personal scores in the personal risk assessment matrix below

PERSONAL RISK ASSESSMENT INDEX		
0 - 5	Low	Total:
6 - 10	Moderate	
11 - 15	High	
16+	Extreme	

FIGURE 2.3 Filled-in example of the ICAS risk assessment matrix (International Council of Air Shows, 2012).

2.3.3.1.12 5M

As discussed earlier, under a zero-tolerance approach, some of the air show performers also identified the 5M model approach as key to risk mitigation (see Figure 2.2). These performers identified the air show industry as a socio-technical system that encompasses a tight coupling of social and technical factors related to the human and their environment with implications for optimal performance, as suggested by Reason (1991) and Dekker (2014). Some of the mentioned mitigations involved the human, the aircraft, the environmental conditions, the display profile itself, and the air show management by event organizers and air bosses. Figure 2.4 illustrates an organized overview of the interconnections and relationships between these elements, emphasizing their significance in a broader context; It visually conveys the intricate web of interactions and dependencies within the 5M framework.

2.3.3.1.1.1 Human Air show performers discussed risks associated with human physiological factors encountered during air shows, such as fatigue, G-induced loss of consciousness (G-LOC), sickness, and skill impairment. In terms of psychological factors, the risk posed by distractions, confidence, emotional state, and self-preservation was highlighted (see Figure 2.4).

2.3.3.1.1.1.1 Physiological Risks Associated with the Human Element of the 5M Theme Fatigue in the form of quality sleep deprivation and human body disorientation due to circadian disruptions is a latent hazard that builds up during the air show periods. Fatigue can have deleterious effects on a pilot's cognitive performance, such as the inability to process sensory cues or retrieve stored information (lapses). Cumulative fatigue throughout an air show period was of much concern for air show performers.

Fatigue risk management techniques from other sectors, such as commercial flight operations, can be applied to the air show community. Examples include coping with sleep deprivation using high-quality earplugs and eyeshades. However, there is currently no documented fatigue risk mitigation system or rules specific to the international air show industry.

Air show performers intimated that fatigue risk management is generally the responsibility of individual air show pilots. Organizers can contribute by providing conducive accommodations for good quality sleep. Synchronizing air show schedules to prevent interference with circadian rhythms is also important. Air show operators emphasized that a good rest before the air show is the responsibility of each display pilot and that performers must ensure their fitness to fly by staying sufficiently rested and safe. Mentoring new air show performers by their teammates is crucial for sharing knowledge about controlling and minimizing fatigue during the performance season.

Other physiological factors such as inadequate body hydration, low blood sugar, sickness, and reduced G tolerance can negatively affect a pilot's performance, increasing the risk of errors and adverse safety events. Fatigue and sickness were mentioned as potential threats to effective aircraft handling capabilities since muscle strength, hand dexterity, and locomotion, which are essential for handling skills, can be impaired.

The findings suggest the importance of air show performers arriving fully prepared and up to date with recent training to showcase not only their flying abilities but also the aircraft's performance and maintenance levels. In terms of improving tolerance to G forces during aerobatic maneuvers, the use of aerobics and muscle-strengthening exercises can be helpful.

2.3.3.1.1.1.2 Psychological Risks Associated with the Human Element of the 5M Theme Psychological risks associated with air show performances include emotional risks, distractions, and performers' confidence levels. Overconfidence can be dangerous, but a healthy degree of confidence is necessary. A balance of healthy risk appreciation and self-awareness is crucial for safe and focused displays.

Distractions can disrupt a performer's mental flow and focus on flying a precise display profile. These can include radio chatter, weather factors, crowds, and family

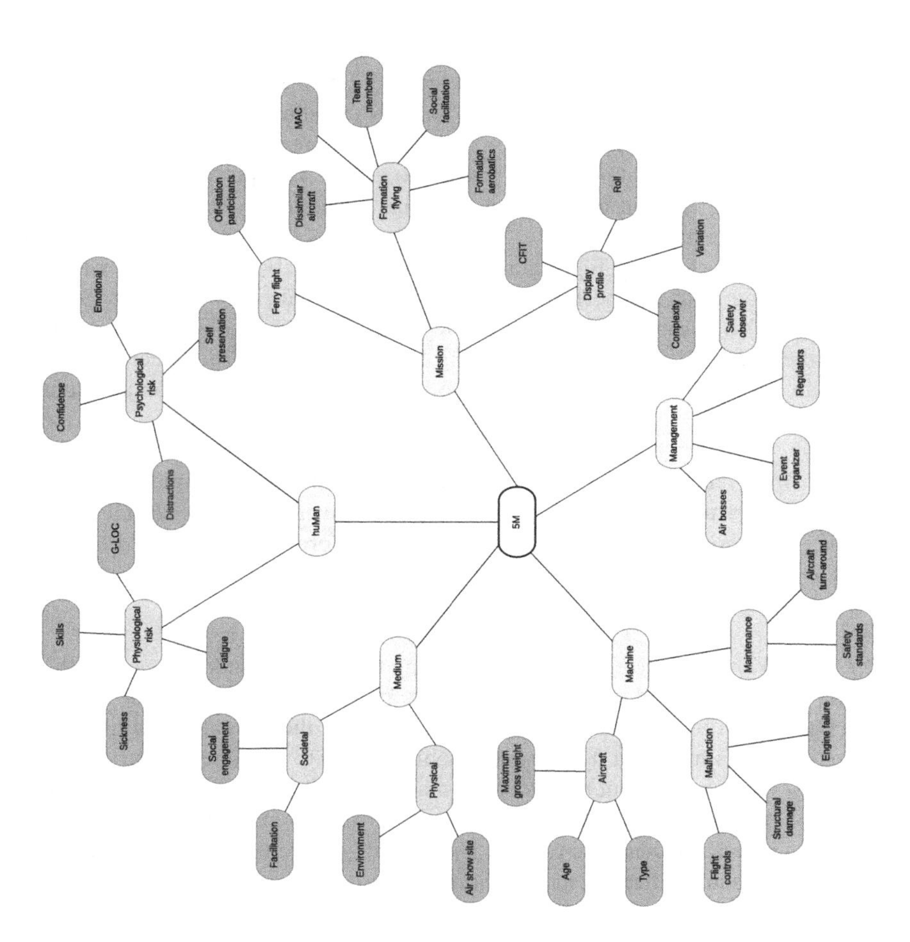

FIGURE 2.4 5M code map.

issues. Emotional hazards such as anxiety, indecision, loss of situational awareness, and stress are also inherent in air show participation. As pilots gain experience, their confidence and emotional stability improve.

Tight time schedules can induce pressure on pilots, causing them to rush through tasks and increasing the potential for errors. Training in mindfulness and meticulous preflight planning can help performers cope better under psychological pressures.

Aircraft handling proficiency and technical skills acquisition was discussed by these air show performers. They focused on a worrisome trend with air show technical and nontechnical knowledge deficit, inadequate experience of some performers, suboptimal training, inadequate planning and preparation, and insufficient currency/recency in air show profiles. The air show performers recommended that organizers ensure pilots have sufficient flight planning resources and preparation time to reduce the feeling of being rushed. Organizers must also ensure that performers are current and meet training and proficiency requirements stipulated by regulators before allowing them to take part in shows.

Some air show performers highlighted the concept of self-preservation, which includes protecting both the performer's life and their financial well-being. Self-preservation is also essential for the air show industry as a whole, as risks that jeopardize the industry can have far-reaching consequences. They perceived this concept as an implicit psychological primer for effective risk appreciation and safe decision-making.

2.3.3.1.1.2 Machine Most air show performers identified aircraft malfunctions, such as engine or structural failures, as potential risks during air shows (see Figure 2.4). While proper maintenance is crucial, air show performers should still be cautious of potential engine malfunctions that could lead to dangerous situations like engine out landings.

2.3.3.1.1.2.1 Warbirds Flying It was interesting to note that under this code, the operational risks associated with flying warbirds or vintage airplanes that are no longer in production were discussed. These risks include the aircraft's age and the challenges of maintaining and flying such machines. Due to the potential structural weaknesses of such obsolete but functional aircraft, warbird operators must be aware of these risks and adapt their flight display profiles to avoid exceeding operational limits. Extra vigilance is necessary when flying vintage aircraft, as they are more prone to engine-related emergencies. Additionally, air show performers, especially warbird pilots, are concerned about the possibility of flight control-related emergencies during displays, as they present a higher risk compared to engine failures.

2.3.3.1.1.3 Management Management of risks in an air show rests mainly with the air boss (see Figure 2.4); nonetheless, the event organizer is accountable for the organization and conduct of the aviation event, especially in the United States (Federal Aviation Administration, 2020a).

2.3.3.1.1.3.1 Air Bosses Air show performers emphasized the importance of the air boss in the tactical management of an air show. The air boss is responsible for the safe execution of all activities at an air show and minimizing operational pressures on-air show performers. They ensure performers stick to the routine and do not introduce new or different maneuvers. Only well-trained, technically competent, and qualified individuals should hold this role as air bosses.

2.3.3.1.1.4 Medium The medium code was undergirded by the physical medium and the societal medium, as per Harris (2011). The physical medium discussed by the stakeholders was related to the air show site complexity, the air show site airspace, obstacles in the vicinity of the area, and the possibility of hitting them (see Figure 2.4). Various environmental factors were also addressed, including marginal weather conditions such as a low cloud ceiling, strong winds, and a high-density altitude; the sun's position in the sky; and bird and drone strikes. Moreover, several air show performers identified that social facilitation bias significantly affected them.

Bird strikes are a primary concern for air show performers, especially for single-engine airplanes, as they pose a higher risk of engine failure. Unauthorized drones in the airspace are also a significant threat, particularly to single-engine jet display teams. High-density altitude affects aircraft performance, resulting in decreased performance, sluggish controls, wider turning radius, and longer display duration. It is also normally associated with low oxygen saturation, high temperatures, and high susceptibility to hypoxia. High-density altitude flying can potentially affect the muscle strength and cognitive abilities of a pilot due to impaired metabolic functions. Mitigation strategies include proper planning and adjusting the display profile as needed.

Flying over water presents environmental hazards, as it can be challenging to gauge distances and altitudes accurately. Pilots need to be cautious, especially when flying over water in the late afternoon with low sun angles, as this can make it even more difficult to perceive their surroundings. Adjusting altitude and avoiding certain maneuvers can help mitigate these risks.

2.3.3.1.1.4.1 Social Risks Social facilitation bias can significantly affect air show performers, causing them to undertake actions that exceed their abilities or skills in order to impress the audience. To minimize this risk, one air show performer intimated his reliance on good training and a strict disciplinary approach to maintaining consistent flying practices, regardless of the crowd size. However, the relationship between air show performers and event organizers can sometimes encourage performers to take unacceptable risks to keep the show exciting and secure their jobs. Air show facilitation might intimidate novice pilots who want to keep their jobs or experienced pilots who want to impress event organizers, who could be personal friends.

2.3.3.1.1.5 Mission Various factors related to the mission were highlighted by air show performers (see Figure 2.4). One risk mentioned is related to ferry flights, which can be dangerous due to unexpected and unanticipated events, even though they are not part of the display mission itself. Some performers also highlighted the risk of colliding with the ground or objects on the ground during air show displays. To mitigate this risk, a fast jet solo display pilot emphasized the importance of knowing the display site, surroundings, obstacles, and building altitudes to maintain a safe margin above the ground.

2.3.3.1.1.5.1 Complexity of the Display Profile Several air show performers mentioned the complexity of the display as a risk factor. One demo team leader emphasized the importance of not relying on high-skill maneuvers, as they cannot guarantee that every team member is fit for every display. The same performer suggested that a display should be set up at a level that requires 80–90% of a pilot's capabilities, leaving a safety margin. Performers should adjust their profiles according to their experience and skills.

Complex missions like circling jumpers or flying in dissimilar aircraft formations can create additional risks. Downline rolls[2], in particular, pose a significant risk to the display profile and could lead to fatal accidents. This supports Barker's (2020b) report, which advises air show performers to exercise extreme caution when incorporating downline rolls into their display profiles.

2.3.3.1.1.5.2 Formation Flying Performers identified several risks associated with formation flying during air shows, with the most prominent risk being midair collisions (MAC). These collisions can occur when routines change or mistakes are made. All team members play a critical role in preventing MACs, and the risk increases if a team member makes a significant error.

The importance of teamwork and clear communication among team members are crucial to avoid incidents. One example given by a performer highlighted the confusion caused by different formation positions based on team members' backgrounds, which led to a near collision. Overall, flying in formation while performing aerobatics increases the risks associated with air show performances.

2.3.3.1.1.5.3 Passenger During a Display Lastly, another risk that could be accepted is the added risk of flying with a passenger during a display. A military air show performer reported offering incentive rides to other air show acts and personnel and using it as a recruiting tool. To reduce the risk, this type of flight is only performed on practice days. No passengers are permitted to be carried in the airplane during the actual display days.

2.3.3.2 Event Observation

To have a more nuanced perspective of the air show display environment, a field observation was conducted by one of the authors during the rehearsal and display days at a southeastern European air show scheduled for the first weekend of September 2021. The author examined operational elements contributing to risks and hazards for air show performers and air bosses.

The operational portion of the event comprised activities that ensured the air displays were conducted safely. The flying control committee (FCC) was responsible for planning, briefing, monitoring, and controlling all ground and flying activities during the air show. The FCC was led by a flying display director, who happened to be the author, assisted by three flying display director assistants, two of whom had extensive military operational experience and the third who specialized in helicopter operations.

The FCC was stationed in the control tower – abeam the show center – giving them a bird's eye view of the display area and all ground movements and activities of the participating aircraft. Additionally, the FCC was assisted by ATC personnel from the local Air Force Base in several aspects of the event's execution. Additionally, a ground managing crew, comprised of two "follow me" cars, supervised all ground activities associated with the event, including aircraft start and taxi, public control, or any other coordination required.

The event featured 15 air show performer entities. The air show performers had varying experiences and engaged in different types of flying activities. These included three international military fast jet solos, two international aerobatic teams, and a formation of two international military fast jets that demonstrated closed circuits and flybys in front of the public.

There were helicopters and fast jets from the three branches of the host nation's armed forces, namely the Air Force, Navy, and Army, which demonstrated tactical flight scenarios in front of the public and also dropped military parachutists. Additionally, the public was allowed to visit two static display areas: one in the west, which included six helicopters from the host Army and Navy, and another in the east, which featured military aircraft from numerous invited Air Forces: Three fast jets and three modern military trainers.

Due to the size of the event and the risk of an incident or accident occurring during the air show, it was necessary to involve various crash and rescue organizations of the host nation. The police, fire brigade, general directorate of protecting civilians, ambulances, and hospitals, as well as the local Civil Aviation Authority, provided not only assistance but also their expertise in preparing and effectively planning the event.

Mission items related to the execution of the air show performances were assessed for each air show performer. Timetables, risk assessment matrices, flight plans, aircrew safety briefings, and weather reports were examined thoroughly to identify any safety-related information applicable to the current study.

2.3.3.2.1 Observation Data Analysis

One of the authors observed numerous aspects of inherent risks and hazards at the event's principal working venues, including the main briefing room, the aircraft parking areas, the control tower, and the crisis and disaster control center.

Moreover, the current air show observation demonstrated how the air show industry could be prepared for potential risks before and during the event, not just from the air show performers' standpoint but also from the organizers', air bosses', and aviation authorities' perspective.

Air show performers observed during both the preparation and execution phases of the event displayed a high level of adaptability to a dynamically changing flying environment, which is a function of resilience, and it was gratifying to see such attributes in the air show industry. For example, constantly changing weather conditions, changes in daylight hours and sun angles, and changes in takeoff and display times due to delays caused by other operational considerations, such as runway inspection for foreign object damage, were among the factors that affected the normal flow of air show performers' display.

During the event observation, both demonstration team leaders noticed winds aloft as a hazard during the rehearsal day. To accept the additional risk, they provided extra buffers to the display lines to ensure the safe execution of their display profile.

Nevertheless, the air show performers' extensive operational expertise enabled them to be resilient and ensure the event's safety. This extensive operating experience validates Hollnagel's (2006) suggestion that safety is not a matter of luck but rather a result of resilience.

2.3.3.2.2 Preflight Risk Assessment Methods

Preflight risk assessment started several months before the event from the air show performers and the air show's FCC team. A display order was developed by the FDD and approved by the local CAA. It was distributed to all air show performers and participants in a summarized version as a "pilot's guide" to assist them in the planning process.

Materials that were also prepared and shared with the air show performers as part of the risk assessment process included a detailed timetable for each day of the event, a risk assessment worksheet for each participant, a safety briefing (see Figure 2.5), and a continuous update on the current Bird/Wildlife Aircraft Strike Hazard (BASH) status and the prevailing and forecasted weather conditions.

During the planning phase, air show performers identified and addressed expected risks. However, the majority of air show performers perceived wind direction as a significant preflight hazard that could affect their display flow, as well as their adherence to the display lines.

Indeed, the researcher's air show site observation was significantly influenced by the implementation of COVID-19 rules and regulations, which affected the event's planning and execution. Crucial pandemic-related measures, such as mask-wearing, social distancing, and avoiding the use of unsanitized objects like papers and pencils during briefings, significantly reshaped the preflight risk assessment strategy for the air show's organizers and the flying control committee. Amid these challenges, innovative strategies were deployed to mitigate the risk of COVID-19 transmission. This included the utilization of QR codes, electronic timetables, display orders, digital signatures for briefing attendance, and virtual briefings via video call platforms like Zoom or WhatsApp, replacing the need for physical gatherings. These adaptations, aimed at reducing virus dissemination during the air show, serve as a valuable case study not only for the air show industry but also for similar mass gatherings, including music concerts and car races, highlighting how such events can be managed safely amid a pandemic.

2.3.3.3 Triangulation of Findings

The implications of the study's data analysis may include modifying the training approaches for air show performers, incorporating risk perception training to revise

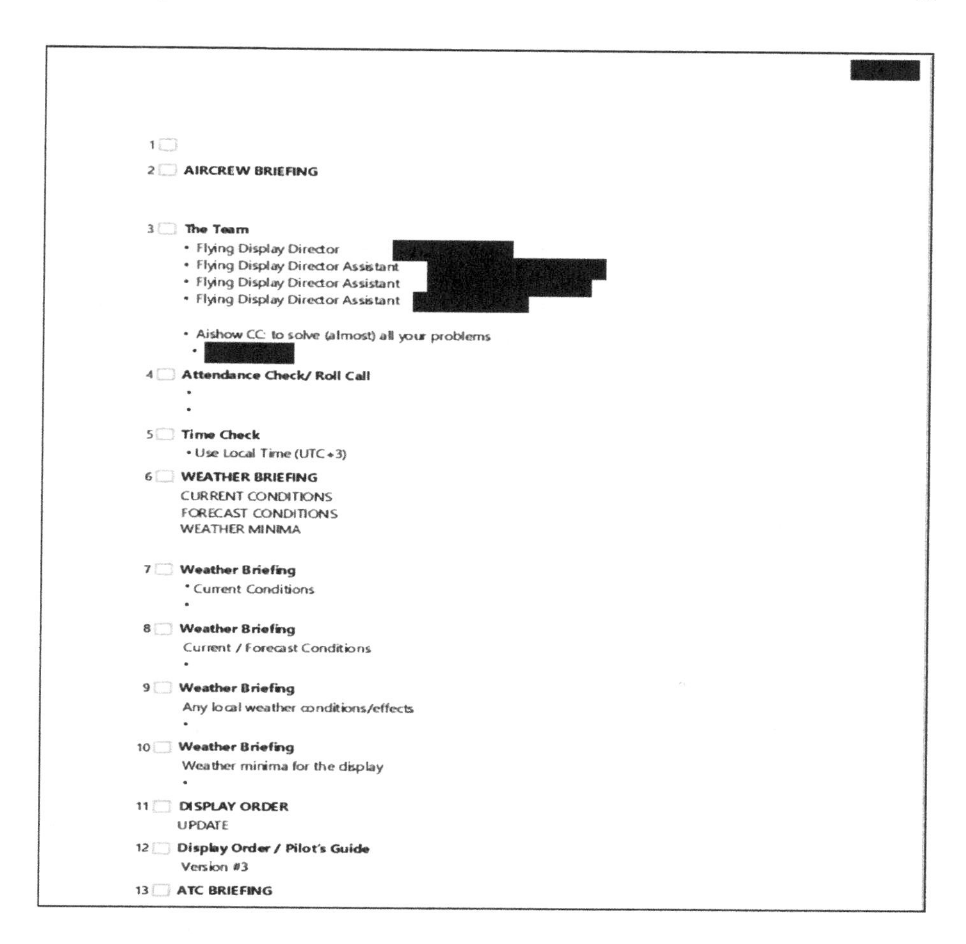

1
2 AIRCREW BRIEFING

3 The Team
- Flying Display Director
- Flying Display Director Assistant
- Flying Display Director Assistant
- Flying Display Director Assistant

- Aishow CC: to solve (almost) all your problems
-

4 Attendance Check/ Roll Call
-
-

5 Time Check
- Use Local Time (UTC+3)

6 WEATHER BRIEFING
CURRENT CONDITIONS
FORECAST CONDITIONS
WEATHER MINIMA

7 Weather Briefing
- Current Conditions
-

8 Weather Briefing
Current / Forecast Conditions
-

9 Weather Briefing
Any local weather conditions/effects
-

10 Weather Briefing
Weather minima for the display
-

11 DISPLAY ORDER
UPDATE

12 Display Order / Pilot's Guide
Version #3

13 ATC BRIEFING

FIGURE 2.5 Air crew safety briefing, an example of briefed items.

the existing safety culture, and even revising the regulations governing aerial events. Especially integrating risk perception training into the training curriculum of air show performers could help them better understand the potential dangers associated with their performances and, ultimately, make informed decisions about the level of risk they are willing to accept.

Specifically, the findings support previous research that found a negative relationship between risk perceptions and higher risk-taking behaviors (Drinkwater & Molesworth, 2010; Ji et al., 2011; Joseph & Reddy, 2013; You et al., 2013). Furthermore, the findings reveal a favorable association between risk perception, mindfulness, and resilient safety culture, confirming the observation that insufficient risk assessment can lead to poor decision-making, culminating in catastrophic aircraft accidents (AAIB, 2017).

Even though there has been a strong focus on safety in the air show community in the past years, unfortunately, fatal accidents still happen, and people lose their lives yearly. The high level of risk involved in air show flying requires an appropriate and relevant level of professionalism and risk management at all levels of administration, from the aviation authorities to the air show organizers, air bosses, and ultimately the air show performers themselves. As suggested by several interviewees, in line with the literature (Barker, 2020a; Chen & Chen, 2014; Schopf et al., 2021), air bosses have the main responsibility to demonstrate effective leadership in managing risks during the planning and execution phase of an aerial event, eliminating the likelihood and severity of any potential risks, while minimizing or zeroing the adverse effects to the public and to the air show industry itself.

2.3.3.3.1 How Do Air Show Performers Perceive and Tolerate Risk Preflight?

Under the section titled "How do Air Show Performers Perceive and Tolerate Risk Preflight," we delve into the enlightening insights and lessons drawn from the observation phase. This practical knowledge contributes significantly to our understanding of risk perception and tolerance among air show performers on the preflight stage.

- Preshow Risk Assessment Planning

Meticulous preshow risk assessment planning emerged as a paramount practice in the successful organization and execution of the observed air show. This involved a comprehensive evaluation of all potential risks and an anticipatory approach to risk management, which commenced as early as 12 months prior to the event.

A significant part of this planning phase was the initiation of communication between the flying display director (FDD) and the air show performers. This communication began several months ahead of the event, establishing a crucial link that facilitated effective coordination and a shared understanding of risks. A major part of this early communication involved a preshow questionnaire to be completed by performers.

This questionnaire served as a tool to gather important information related to performers' needs and considerations for their routines. It allowed performers to make any requests related to their performance, which could range from equipment requirements to specific timing preferences. The questionnaire also prompted performers to highlight any unique risks associated with their performance, ensuring these factors are incorporated into the broader risk management strategy.

This proactive identification of potential risks covers a wide spectrum. It could be equipment-related, such as possible malfunctions or technical issues; weather-related, including risks associated with inclement conditions; or even health-related, pertaining to potential health concerns of the performers. By cataloging these diverse risks, FDDs and performers can collectively develop tailored mitigation strategies for each identified risk.

These mitigation strategies form a crucial blueprint for the successful execution of the air show. They guide actions and decisions in the face of unforeseen circumstances, ensuring a swift, organized response to any issue that may arise. Moreover, these strategies foster a safety culture where risk awareness and preventive action become a natural part of the air show's operation.

In essence, the lessons from the observation phase underscore the critical role of meticulous preshow risk assessment planning. The early initiation of dialogue between the FDD and performers, the comprehensive identification of potential risks via preshow questionnaires, and the proactive development of tailored mitigation strategies form the cornerstone of this planning. This comprehensive, anticipatory approach significantly contributes to the safety and success of air shows.

- Event Timeline

The construction of a precise timeline for all events emerged as a fundamental part of the air show planning and execution during the observation phase. This timeline, calculated down to the minute, provided a comprehensive schedule for each act, minimizing uncertainty and guaranteeing the seamless execution of the show.

This specific event itinerary forms a backbone for air shows, creating a framework that organizes all associated activities. It provides a roadmap for performers, allowing them to precisely plan and time their routines. This level of detail assists performers in managing their energy, reducing the risk of errors due to miscalculated time allowances. Moreover, it affords them the ability to fine-tune their performances, ensuring that their acts are delivered with maximum precision and impact.

Additionally, this highly accurate timeline is invaluable for the ground support team. It allows for exact resource and personnel allocation, optimizing efficiency and ensuring safety procedures are effectively implemented. With an understanding of when each act will occur, ground support can prepare necessary equipment, clear the airspace, and ready emergency services as required. This level of detail can drastically improve the effectiveness of safety protocols and ensure a well-coordinated response to any situation.

The audience also benefits from this precision. With a clearly defined schedule, they can anticipate and appreciate each act fully, enhancing their overall experience. Moreover, a predictable timeline can assist in crowd management, as attendees are better able to plan their movements and activities throughout the show.

- ICAS Individual Risk Management/Assessment Worksheet

The utility of the International Council of Air Shows (ICAS) Individual Risk Management/Assessment Worksheet was another important revelation. This tool provides a tangible way to visualize, categorize, and ultimately address risks, facilitating preemptive action based on the severity and likelihood of the risk involved.

The ICAS Individual Risk Management Worksheet proves its worth by providing an efficient, user-friendly, and comprehensive risk assessment tool specifically designed for air show performers. With its intuitive layout and focused content, it facilitates an effective risk evaluation without imposing an undue time burden on the user.

The worksheet starts with the Scheduling Assessment, which utilizes the IMSAFE checklist (Federal Aviation Administration, 2022) – an acronym for Illness, Medication, Stress, Alcohol, Fatigue, and Emotion. This quick yet vital self-check enables pilots to ascertain their fitness for flight in a streamlined and easy-to-understand manner.

Next, the Preperformance Assessment section evaluates environmental and execution factors. These include a range of elements, from weather conditions to flight sequence details. Designed to be straightforward and pilot-centric, these sections allow for a rapid yet thorough review of external and performance-specific factors.

The Risk Assessment Criteria then prompt pilots to reflect on their personal readiness through intuitive questions, such as "How do you feel?" "Quality of crew rest," and "Days since the last air show." The simplicity and directness of these questions facilitate honest self-assessment without consuming valuable preshow time.

The final section, the Individual Risk Assessment, encourages pilots to consider their personal risk tolerance and decision-making, promoting an individualized approach to risk management. This component is designed to be quick yet effective, allowing pilots to gauge their comfort with the upcoming performance without requiring extensive time commitment.

In summary, the ICAS Individual Risk Management/Assessment Worksheet is an efficient, user-friendly tool that simplifies the complex process of risk assessment. It reduces the time and complexity typically associated with safety checks while still providing a comprehensive overview of potential risks, making it an invaluable asset for air show performers. Its pilot-centric design ensures that, despite the comprehensive nature of the checks it entails, it remains accessible and convenient to use, ultimately contributing to an overall safer environment for air shows.

- Preshow Safety Briefing

The importance of conducting a comprehensive preshow safety briefing session emerged as a pivotal strategy during the observation phase. This meeting serves as an essential platform to disseminate information, clarify responsibilities, and ensure a unified understanding of safety protocols among all involved parties.

In a typical preshow safety briefing, a thorough review of the risk assessment findings is conducted. This review allows every participant to fully understand the potential hazards identified, their implications, and the mitigation strategies designed to handle them. By addressing each identified risk, the briefing ensures that all performers are well-versed in what to anticipate and how to respond, enhancing the collective readiness to manage unforeseen situations.

The safety briefing also includes a detailed walk-through of the event's timeline. A clear understanding of the sequence of activities reduces uncertainty and provides a shared mental model for the day's operations. It not only coordinates performers but also enables the ground support team to anticipate their responsibilities.

In addition to discussing safety protocols and procedures, the briefing offers an opportunity for all participants to ask questions, share concerns, or provide additional insights. This interactive dialogue fosters a sense of shared responsibility, reinforces the team's unity, and empowers individuals to act proactively in the interest of safety.

Moreover, a comprehensive preshow safety briefing underscores the importance of individual roles in ensuring the overall safety of the event. Each team member, from performers to ground crew and air traffic controllers, has a unique contribution to make to the safe running of the air show. A clear understanding of their roles and responsibilities, as well as how they interconnect with others, is integral to the success of the event.

Lastly, these preshow briefings also help set the tone for the air show, reinforcing a culture of safety where risks are acknowledged, managed, and mitigated. They serve as a potent reminder of the shared commitment to safety, fostering a collective mindset where safety is prioritized above all else.

In conclusion, the preshow safety briefing stands as a vital mechanism in the arsenal of risk management tools. It not only imparts critical information but also nurtures a shared safety culture, thereby contributing significantly to the overall safety of air shows.

A standout role observed during the shows was that of the "Air Boss," who, from the control tower, oversees all flight operations. Their integral role in monitoring and managing any arising risks reaffirms the necessity for a dedicated individual to ensure the safe progression of aerial activities.

- Safety Observers

Safety observers' role, particularly for military demonstration teams, proved to be a significant aspect of augmenting safety during the observed air show. They serve as an additional safeguard, a "wingman on the ground" providing an extra layer of scrutiny that substantially contributes to the smooth execution of complex aerial maneuvers.

Safety observers often include seasoned pilots or experienced air show personnel who are intimately familiar with the intricacies of flight operations and air show procedures. They are strategically positioned in line with the show center either on the ground or on the tower to maintain a vigilant watch over the ongoing performance.

One of the primary duties of a safety observer is to continuously monitor the airspace in which the air show performer is operating. They keep a keen eye on the designated airspace for any possible incursions or potential hazards that the performer might not be aware of. By doing so, they provide a crucial safety net, especially in congested airspaces or complex air shows involving multiple performers.

Besides monitoring the airspace, safety observers are also tasked with observing the accuracy of the performance. They compare the actual maneuvers being executed with the planned display sequence to identify any deviations. If their fellow performer deviates from the agreed flight path or makes an unanticipated maneuver, the safety observer can quickly alert them, aiding in the immediate rectification of the error.

In military demonstrations, where the complexity of maneuvers and the proximity between aircraft increase the potential for risk, safety observers play an even more crucial role. Their experienced eyes can quickly detect any issues or irregularities, such as too-close formations or wrong entry speeds, which can then be promptly addressed.

Safety observers also serve as a vital link to the ground crew and air traffic control. They can swiftly communicate any necessary adjustments to the performance, initiate emergency procedures, or call for aborting a maneuver if the situation warrants.

In essence, the role of safety observers in air shows, particularly in military demonstrations, cannot be overstated. They are an indispensable component of the overall safety strategy, providing a vigilant watch over proceedings, alerting teams to potential risks, and facilitating quick responses to evolving situations. They uphold the commitment to safety while allowing the performers to deliver thrilling displays that lie at the heart of every air show.

Taken together, these lessons underline the complex and multifaceted nature of risk management at air shows. Understanding how performers perceive and tolerate risk before the flight provides us with a unique vantage point to improve safety protocols, thereby enhancing the exhilarating yet secure experience that air shows aim to provide.

2.3.3.3.2 How do Air Show Performers Perceive And Tolerate Risk In-flight?

The process of assessing and managing risk during flight by air show performers is a complex interplay of acquired knowledge, practical experience, proficient use of flight instrumentation, and a deep reliance on personal sensory inputs. The crucial role of these elements is most evident in high-stakes environments like air shows where precision, safety, and spectacle need to coalesce perfectly.

Pilots rely on an array of flight instruments and modern display technologies in the cockpit. These tools, ranging from traditional gauges such as the altimeter, airspeed indicator, and artificial horizon to advanced systems like the head-up display (HUD) and flight path marker (FPM), serve to provide essential real-time data about the aircraft's performance and attitude.

HUDs, in particular, are vital tools for air show performers. These systems project critical flight information onto a transparent display positioned in front of the pilot. This allows pilots to monitor necessary data without having to look down at the instrument panel, thus maintaining visual contact with the surroundings. Such information includes airspeed, altitude, heading, and flight path information, contributing greatly to the safety and accuracy of the performance.

The FPM, an integral part of the HUD, indicates the aircraft's actual flight path, showing where the aircraft is headed in three dimensions. It takes into account factors like aircraft attitude, angle of attack, and sideslip angle. For performers executing complex aerobatic maneuvers, the FPM is invaluable in maintaining control and awareness of their aircraft's trajectory.

In addition to the HUD and FPM, the advent of advanced mapping display technologies has significantly enhanced situational awareness for pilots. These systems present an interactive map of the display area, complete with details about nearby obstacles, the predefined display path, and even real-time weather conditions. The enhanced understanding of the display environment that these tools offer allows pilots to better anticipate and respond to potential risks, thus increasing the overall safety of the performance.

However, sophisticated instruments alone aren't sufficient. Pilots also heavily rely on their sensory inputs, particularly peripheral vision and proprioception.

Peripheral vision is often relied upon to maintain situational awareness, especially during close formation flying, where pilots need to maintain a specific relative position to other aircraft. It also aids in spotting potential hazards such as birds, drones, or tall masts of sailboats in overwater displays.

Further, peripheral vision contributes to height perception, providing essential cues about the aircraft's proximity to the ground – a key element in low-level aerobatics. Proprioception, the intuitive feeling of the aircraft's reactions to control inputs, complements this sensory suite and provides a direct connection to the aircraft, allowing pilots to feel the aircraft's response to their control inputs.

The mental processes underpinning decision-making and risk assessment are significantly shaped by the level of training and experience of the performers. Comprehensive training imparts the knowledge and skills necessary to handle unexpected situations and adjust to environmental changes, while accumulated experience heightens the performer's instinctual capacity to evaluate situations and decide swiftly and accurately.

Indeed, the complexity of a display correlates directly with the level of risk and the mental strain imposed on the performer. However, with stringent practice, strict adherence to a pre-planned display profile, support from the ground crew, and sensory acuity, performers can effectively manage this complexity, executing their displays safely and efficiently.

Finally, it's important to emphasize that experience plays a pivotal role in risk tolerance. The more experienced performers are, the better their skills in anticipating, reacting, and adapting to rapidly changing situations. Through rigorous training and accrued flight experience, performers develop the mental readiness required to manage the complexities and inherent as well as unexpected risks associated with air show performances.

2.3.4 Variations in Risk Management Based on Demographics

During the survey conducted as part of the study, the following demographic groups were represented in the responses: (a) age; (b) gender and country of origin; (c) marital status and educational background; and (d) current role in the air show community, air show flying experience, and aerobatics background. In addition to Canada,

France, the United Kingdom, and the United States, the respondents originated from the following seventeen countries: Belgium, Chile, the Czech Republic, Denmark, Finland, Germany, Ghana, Greece, Italy, Jordan, the Netherlands, Norway, Poland, Romania, South Africa, Sweden, and Switzerland.

The analysis included an examination of the relationships between resilient safety culture, safety risk parameters, and mindfulness in the international air show community using air show flying experience, military or civilian flying experience, age, educational background, and marital status as demographic variables.

a. Air Show Flying Experience

The results reveal a positive correlation between air show flying experience and risk perception, supporting prior research that establishes a clear positive link between experience, be it in flying or other activities, and risk perception (Crundall et al., 2013; Ferraro et al., 2015; Joseph & Reddy, 2013; Winter et al., 2019; You et al., 2013). The data implies that as air show pilots gain more experience, they may receive additional training and develop a deeper comprehension of the aircraft's dynamics and energy management. A pilot performing in their first air show may have limited understanding of unforeseen risks, as their risk perception is solely based on information obtained from training materials, lessons shared by mentors and instructors, and minimal exposure to low-altitude aerobatic environments.

Inexperienced performers may still be unaware of certain hazards and risks that have contributed to past air show incidents. As Bob Hoover (Barker, 2020b) pointed out, "There are no new accidents, only new pilots causing old accidents." Continuous education through mentorship may help air show performers bridge gaps in experience, as recommended by ICAS (2023) and the UK CAA (2023a). However, the results also demonstrate a negative correlation between air show flying experience and risk tolerance, reinforcing Barker's (2003, 2020a) claim that, unlike in general aviation, an air show pilot's flying experience does not guarantee a safe performance.

Furthermore, several accident investigations have found that highly skilled air show pilots, particularly when performing solo, may challenge themselves and their aircraft to the extreme, leaving no room for error (BFU Switzerland, 2005; National Transport Safety Board (NTSB), 2011; UK Air Accidents Investigation Branch, 2017).

b. Military or Civilian Flying Experience

The findings of this study show no significant correlation between a military or civilian background and risk perception and tolerance. However, there is a notable difference in the perceived usefulness of safety risk matrices between military and civilian air show performers. Most civilian air show performers who were interviewed expressed that risk assessment matrices do not add value to their work, as they see them as part of a bureaucratic process. Considering that a large number of air show performers fly solo and act as a one-person organization, handling everything from administration to display training and aircraft maintenance, they may regard an SMS or risk matrix as unnecessary paperwork that diverts their focus from achieving their goals. Nevertheless, by adopting and adapting specific sections of an SMS, as recommended by the FAA (2020b), civilian air show performers could tailor these systems to suit their operations, ultimately improving the safety of their air displays.

c. Age

The data analysis reveals a positive connection between age and risk perception. Conversely, the findings indicate a negative correlation between age and risk tolerance, aligning with previous research (Gibson et al., 2013; Hallahan et al., 2004) that suggests an inverse relationship between age and risk tolerance. As pilots grow older, they tend to become more mature in various aspects of life and gain a heightened awareness of the inherent risks they encounter daily, not only in flying but also in general life experiences. Consequently, they may be more inclined to accept lower levels of risk.

However, age-related physiological factors, such as heart diseases, which are considered in medical certification for pilots (National Transport Safety Board (NTSB), 2011), could also impact the physiological state and overall risk assessment of air show performers. Nonetheless, the examination of this risk factor was not within the scope of the study.

d. Educational Background

The study's findings reveal a positive correlation between educational background and risk perception while also indicating a negative correlation between educational background and risk tolerance. This supports Chionis and Karanikas' (2018) assertion that aviation professionals with postgraduate degrees tend to be less risk-averse than their counterparts who hold a bachelor's degree or lower.

Air show performers with a master's degree or higher have been exposed to a more advanced level of education and knowledge, which may not only enhance their understanding of risk assessment processes and safety theories but also provide them with the maturity to comprehend their responsibilities toward aviation authorities and, more importantly, the air show community.

However, this does not imply that air show performers with lower education levels are unsuitable for the air community because they might be more risk-averse. Instead, pilots with lower educational backgrounds may require additional support from regulators, mentors, and their peers to ensure they possess the necessary risk management skills.

Supplemental educational activities at air show conventions, such as workshops, seminars, or webinars, as well as mentorship from more experienced display pilots, could benefit pilots with lower educational backgrounds.

e. Marital Status

The study found that being married positively correlates with risk perception and negatively correlates with risk tolerance, aligning with previous research findings

(Aumeboonsuke & Caplanova, 2021; Hallahan et al., 2004) that identify marital status as a crucial factor affecting risk tolerance scores.

Given the relationship between marital status and risk factors, it is important to focus on mental preparedness training (Andersen et al., 2016) for air show performers who may be influenced by their marital status (Scheid & Wright, 2017). Lessons learned from air show performers preoccupied with family concerns could be discussed during annual ICAS and EAC air show safety workshops, with mitigation strategies being provided. Additionally, based on the findings, support groups for unmarried (single) air show performers could be established to help reduce their risk tolerance and improve their risk perception.

Married pilots' preoccupation with family issues may affect their mindfulness and increase their risk tolerance. Although married air show performers might have more stable lives compared to single performers, concerns involving family members, finances, and other matters could distract them before a flight. Notably, the ICAS individual risk assessment worksheet (International Council of Air Shows, n.d.) already includes a risk criterion for air show performers to report how many months they are expecting a child. This shows that the air show community has already identified and addressed family-related risk factors.

2.3.5 Practical Examples of Effective Risk Management Methods

Numerous examples of successful risk management methods across the air show community exist. Both air show performers and flight control committees conduct a thorough risk assessment of the hazards affecting their air shows, yet below it will be discussed the risk management methods followed by one civilian air show performer, one military air show performer, and one civilian air show organization. That would be Patty Wagstaff, the Royal Air Force Red Arrows aerobatic team, and the Athens Flying Week International Air Show, respectively.

One successful example of risk management methods used by a civilian air show performer Patty Wagstaff. Wagstaff is a world-renowned aerobatic pilot known for her precise and thrilling performances. Like other air show performers, she employs a comprehensive risk management plan to ensure her safety during performances. Wagstaff's risk management methods include preflight briefings, safety equipment requirements, safety checks, weather monitoring, and emphasis on emergency response teams (Wagstaff, 2016). She also stresses the importance of maintaining physical fitness and undergoing regular medical checkups to ensure that she is fit to perform.

Before every performance, Wagstaff and her team analyze the flight plan, aircraft performance, weather conditions, and other critical factors that could affect safety. This analysis allows them to identify potential hazards and develop effective risk mitigation strategies. Additionally, Wagstaff undergoes regular training to ensure she keeps herself physically conditioned to G-forces while remaining familiar with safety protocols, emergency procedures, and other critical information.

Then, the Red Arrows, the aerobatic display team of the Royal Air Force (RAF) in the United Kingdom, demonstrate another successful example of risk management methods. The team is known for its precision flying and intricate formations. The Red Arrows employ a comprehensive risk management plan to ensure their safety during performances. The team undergoes extensive training and participates in regular practice sessions to ensure they can work together seamlessly during performances. Effective communication between team members, ground crews, and air traffic controllers is also crucial (Royal Air Force, United Kingdom, 2023)

A prime example of air show organizers implementing exceptional risk management practices is the Athens Flying Week (AFW) International Air Show in Greece. This prestigious event showcases the latest innovations in aviation and draws thousands of spectators annually (Athens Flying Week International Air Show (AFW), 2023). The successful organization and execution of such a large-scale event necessitate efficient risk management strategies to guarantee the safety of participants, spectators, and the environment. This involves starting with a blank sheet every year and employing various risk assessment tools, such as the bowtie risk assessment tool (UK Civil Aviation Authority, 2023b) and a risk level timeline displayed in a dashboard format.

Initiating with a blank sheet each year is crucial to identify and evaluate all potential hazards. This approach averts complacency and assists organizers in considering new risks that may have emerged since the previous event. Moreover, it ensures that the risk management process is continuously updated, accounting for changes in technology, regulations, and other external factors that might influence the event.

All in all, effective risk management is critical to ensuring the safety of air show performers. By implementing comprehensive risk management plans that cover key components such as risk assessment, risk mitigation, emergency response planning, communication, training and education, and safety checks, air show performers can minimize the risks associated with their performances. Examples such as Patty Wagstaff, the Red Arrows, and the Athens Flying Week International Air Show demonstrate that successful risk management plans can be employed to ensure performers' safety and the safety of all people in and around the premises of the air show.

2.4 ENHANCING RISK MANAGEMENT IN AIR SHOW PERFORMANCES

In this chapter, we will explore three additional subjects that lay the groundwork for improved risk management among air show performers. We will start by introducing a new operational risk management framework specifically designed for air show performers, along with a suggested acronym. Following that, we will examine a methodical approach to risk management before concluding with a discussion on a hidden yet significant risk to air shows. The final topic has been purposefully placed at the end of the chapter to leave a lasting impression on readers.

2.4.1 Airshow Operational Risk Management (AORM) Framework for Performers

Airshow operational risk management (*AORM*) is a systematic process of identifying, assessing, and managing risks associated with air show performances. It helps air show organizers, performers, and other stakeholders make informed decisions to reduce potential hazards and improve safety. The *AORM* framework aims to improve safety and minimize accidents at air shows by recognizing, evaluating, and controlling risks. Drawing on industry best practices and the authors' recommendations, the *AORM* concept is tailored to address air show-specific concerns.

Integrating an *AORM* in preshow planning is vital for identifying and mitigating all associated risks while ensuring the safety of everyone involved. In the context of this book's current edition, a practical 10-step ORM framework for air show performers will be provided, complete with an easy-to-remember acronym: *SHOWCENTER*© (see Figure 2.6). By keeping the *SHOWCENTER*© acronym in mind, air show performers can follow this 10-step ORM framework to improve safety, reduce risks, and guarantee a successful performance.

2.4.1.1 Reasoning for the SHOWCENTER© Acronym

The acronym *SHOWCENTER*© was deliberately chosen after careful deliberation to make it memorable for air show performers. The showing center in an air show is a critical reference point for performers, air bosses, and spectators. It serves as the focal point of the aerial displays and is where the majority of the aerial maneuvers are executed. The importance of the show center lies in the following aspects:

- *Coordination and Orientation*

The show center helps pilots to maintain their orientation in relation to the airfield, audience, and other performers. It enables them to accurately execute their routines while ensuring proper spacing and coordination with other aircraft. For air show performers, the show center is also vital for coordination with the air boss and ground-based safety observers. In some demo teams, a mandatory "show center" call by the safety observer helps pilots gauge their performance, allowing them to make necessary adjustments. Additionally, it serves as the primary point for drawing display lines, offset lines, and abort lines in display profiles.

- *Safety*

By having a clearly defined show center, pilots can maintain a safe distance from the spectators and avoid any potential hazards on the ground. This minimizes the risk of accidents and ensures the safety of both performers and the audience.

- *Visibility*

The show center ensures that the aerial performances are optimally visible to the audience. Pilots plan their maneuvers around this central point, allowing spectators to enjoy the displays without straining to see distant or obscured action. As a

S	**S**urvey the situation: Gather information about the air show's specific conditions, such as weather, location, and potential hazards.
H	**H**azard identification: Identify any potential risks or hazards related to the performance, including aircraft limitations, pilot health, and the complexity of the maneuvers.
O	**O**bjective assessment: Define the goals and objectives of the performance, ensuring they align with the pilot's capabilities and the aircraft's limitations.
W	**W**eigh alternatives: Explore different options for addressing the identified risks, including modifying the performance, adjusting maneuvers, or implementing additional safety measures.
C	**C**ompare risk levels: Evaluate and compare the potential risks associated with each alternative, taking into account the probability and severity of potential incidents.
E	**E**stablish controls: Develop and implement appropriate risk controls, such as predefined abort points, communication protocols, or backup plans.
N	**N**avigate contingencies: Prepare for unexpected events by developing contingency plans and practicing emergency procedures.
T	**T**rack performance: Continuously monitor the performance during the air show, paying close attention to any changes in conditions or the emergence of new risks.
E	**E**valuate outcomes: After the air show, review the performance and the effectiveness of the implemented risk controls, identifying any areas for improvement.
R	**R**efine the process: Based on the evaluation, make adjustments to the AORM process to enhance safety and performance in future air shows.

FIGURE 2.6 *SHOWCENTER* acronym.

common practice, it is situated in front of the main VIP area, making it the focal point during maneuvers.

• *Timing and Flow*

The show center aids in managing the timing and flow of performances. By using it as a reference point, pilots can smoothly transition from one maneuver to another and maintain a seamless flow of aerial displays.

Considering all the aforementioned factors, it appears that the show center plays a pivotal role for air show performers, as everything begins and concludes at this focal point. When display pilots hear the term "show center," various concepts spring to mind. This performer-centric acronym is useful not only before and after air shows but also during the events, allowing pilots to conduct self-checks and risk assessments as needed.

The *SHOWCENTER©* acronym encompasses both safety and excellence aspects, encouraging continuous assessment of risks and performance both preshow and during the show. It promotes resilience, adaptability, risk management, and excellence for a safe and impressive display. The hope is for the acronym to become relevant not only to air show performers but to all aviators, aiding in self-assessment, focus, and mindfulness throughout their flights.

Furthermore, the *SHOWCENTER©* framework can be extended to daily life, helping people self-assess, improve, and stay focused on their goals. It serves as a reminder to stay mindful and on track, seeking solutions to return to one's "center" when faced with challenges.

While primarily conceived for air show performers, the *SHOWCENTER©* framework can be employed by anyone, regardless of their connection to aviation. It represents the central focus that individuals should maintain at all times. If one finds themselves distracted or off course, they must return to their "show center," reestablishing focus on their primary goals. This concept also highlights the resilience of individuals who can refocus and return to their primary tasks.

2.4.1.2 Risk Scores, Levels, and Relevant Mitigation Strategies

As part of the *AORM*, a practical method to effectively manage risks during an air show is to use a formula for calculating risk scores and applying the resulting scores to a risk matrix. This process helps to categorize risks and identify appropriate mitigation strategies. Here's a step-by-step guide on how to use the risk score formula, risk matrix, and corresponding mitigation strategies (see Figure 2.7):

- *Identify Hazards*

An air show performer must conduct a thorough risk assessment and identify any potential hazards that may be encountered during an air show. While looking for information on the air show from the air boss, previous personal experiences, and peer support, hazards could be identified using the *SHOWCENTER©* acronym.

These hazards may include, but are not limited to:

- Weather conditions (e.g., strong winds, turbulence, low visibility)
- Aircraft mechanical issues or failures
- Mid-air collision
- Loss of situational awareness
- Pilot health issues (e.g., fatigue, disorientation)
- Ground incidents (e.g., runway incursions, accidents)
- Fire or fuel-related hazards
- Wildlife hazards (e.g., bird strikes)

- Assess probability and severity

For each potential hazard, assess the probability (likelihood of occurrence) and severity (the impact of the hazard if it occurs). Assign numerical values for probability and severity based on predefined scales (e.g., 1 to 5, with 1 being the lowest and 5 being the highest).

An example of a risk matrix with probability and severity scores is depicted in Figure 2.8.

- *Calculate Risk Scores*

Use the risk score formula (*Risk Score = Probability x Severity*) to calculate the risk score for each hazard by multiplying the assigned probability and severity values.

- *Categorize Risks*

Based on the risk scores, categorize the potential hazards into predefined risk levels (e.g., low, moderate, high, very high, critical). This categorization allows for prioritizing risks and determining appropriate mitigation strategies.

Figure 2.9 shows a risk matrix that combines probability and severity values, resulting in risk scores. The matrix should provide a visual representation of the relationship between probability and severity, making it easier to identify, categorize, and prioritize risks.

- *Designate Mitigation Strategies*

For each risk level, designate appropriate mitigation strategies that address the specific risks associated with that category. These strategies should be tailored to the air show performers and the event's unique circumstances.

Designating risk levels and assigning risk scores can help air show performers prioritize risks and identify appropriate mitigation measures during the planning phase. The risk levels can be categorized as low, moderate, high, very high, and critical. Table 2.2 depicts a breakdown of risk scores and risk levels, along with suggested mitigation measures and practical examples for each risk level:

In Table 2.2, risks are categorized based on their scores, with corresponding mitigation strategies tailored to air show performers. By implementing these strategies, performers can address potential hazards and enhance safety during their performances.

- *Implement Mitigation Strategies*

Apply the designated mitigation strategies to address and minimize the risks identified. This process may involve training, communication, contingency planning, or other actions depending on the specific risk level and hazard.

2.4.1.3 Examples

In Table 2.3, three examples of hazards are presented with varying risk scores, from low to critical. The associated mitigation strategies are designed to address each

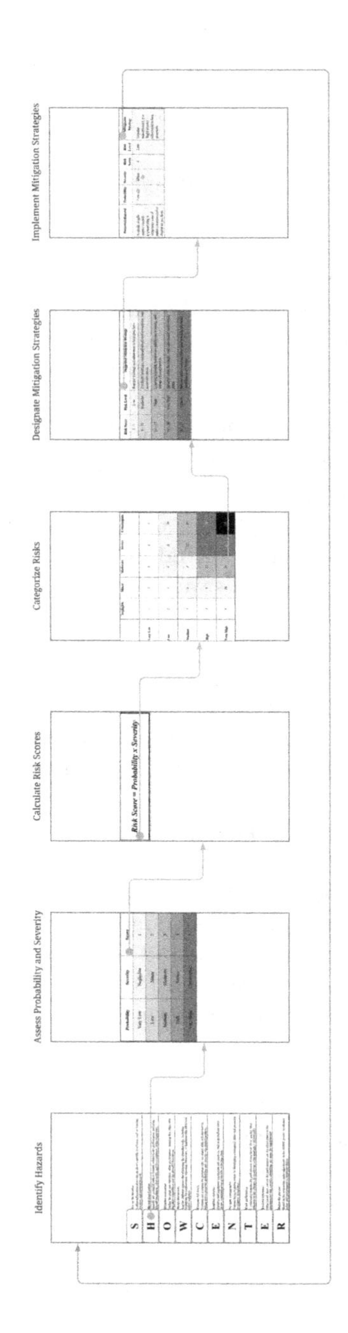

FIGURE 2.7 Guide to risk mitigation strategies.

Probability	Severity	Score
Very Low	Negligible	1
Low	Minor	2
Medium	Moderate	3
High	Severe	4
Very High	Catastrophic	5

FIGURE 2.8 Risk matrix with probability and severity scores.

specific risk category, helping air show performers and organizers reduce potential hazards and enhance the overall safety of their events.

By understanding the risk levels, scores, and appropriate mitigation measures, air show performers can effectively prioritize and manage risks, promoting safety and excellence. The *AORM* framework will enable better decision-making and resource allocation while reducing potential hazards and ensuring the safety of both performers and spectators.

2.4.2 Systematic Approach to Risk Management in Air Shows

While human error is often cited as a significant contributing factor in air show accidents, it is essential to recognize that attributing incidents solely to air show performers may not be entirely fair. Incidents, accidents, or near misses resulting from human actions often stem from systemic issues that develop over time. It is crucial to differentiate between risks within human control and external factors beyond human control.

Moreover, accidents caused by an air show performer's incorrect actions may be the culmination of a chain reaction involving multiple errors or a series of broken links in the chain of events. As a result, a continuous, dynamic, adaptable, systemic, and systematic risk management approach is necessary, starting from the initial training stages for an air show performer.

It is vital not to leave anything to chance or place the blame solely on the pilot or air show performer when things don't go as planned. A systematic approach involving all air show stakeholders is crucial, encompassing air show performers, air bosses, regulators, air show organizers, and the entire air show community. Working as a team to ensure effective communication and address all risks associated with air shows is essential. By identifying proper mitigation strategies and speaking up when noticing deviations from established standards, stakeholders can collaboratively take

	Negligible	Minor	Moderate	Severe	Catastrophic
Very Low	*1*	*2*	*3*	*4*	*5*
Low	*2*	*4*	*6*	*8*	*10*
Medium	*3*	*6*	*9*	*12*	*15*
High	*4*	*8*	*12*	*16*	*20*
Very High	*5*	*10*	*15*	*20*	*25*

FIGURE 2.9 Risk matrix with the relationship between probability and severity scores.

TABLE 2.2
Breakdown of Risk Scores and Risk Levels, along with Suggested Mitigation Measures and Practical Examples

Risk Score	Risk Level	Mitigation Strategy
1–5	Low	Regular training and adherence to best practices.
6–10	Moderate	Preflight briefings, increased situational awareness, and communication.
11–15	High	Limiting high-risk maneuvers, additional training, and enhanced coordination.
16–20	Very High	Detailed safety briefings, risk assessment, and backup plans.
21–25	Critical	Review and reassess air show performance, potential cancellation of display.

TABLE 2.3
Practical Examples of Air Show Hazards With Suggested Mitigation Strategies

Potential Hazard	Probability	Severity	Risk Score	Risk Level	Mitigation Strategy
A small, single-engine aircraft experiencing a temporary loss of radio communication during an air show	Low (2)	Minor (2)	4	Low	Regular maintenance, preflight checks, adherence to best practices
An air show is scheduled to take place near an active wildlife park; possible bird strike	High (4)	Severe (4)	16	Very High	Mandatory briefings, close monitoring of bird activity, abort maneuver training, enhanced communication
Engine failure during display	Very High (5)	Catastrophic (5)	25	Critical	Review performance, rigorous inspections, contingency plans, and consider canceling the display

corrective action to maintain safety. As such, accident investigation reports should consider various factors and systemic issues rather than focusing only on individual errors.

Achieving a systematic approach to risk management in the air show industry requires concerted efforts from all stakeholders. Part of this effort includes developing a safety management system (SMS) that addresses hazard identification, risk assessment, and mitigation strategies specifically for air show organizations. According to the International Civil Aviation Organization (ICAO), an SMS is defined as "a systematic approach to managing safety, including the necessary organizational structures, accountabilities, policies, and procedures" (ICAO, 2018, p. viii). However, creating a full-scale SMS can be resource intensive, and for some air show performers or organizations, the financial burden may be a significant obstacle. In such cases, implementing an *Airshow-SMS* can provide a more sustainable solution for the industry, offering a cost-effective approach to risk management while still promoting safety and reducing accidents.

2.4.2.1 Airshow-SMS: An Option for Smaller Organizations and Single-Person Air Show Performers

In the realm of aviation, the safety management system (SMS) stands as a globally recognized framework for managing safety (ICAO, 2018). However, its implementation often brings to mind large operations and commercial airlines. In this section of the book, we address this misconception and introduce the concept of the "Airshow-Safety Management System" (*ASMS*), a scaled-down version of the SMS, adaptable and effective for individual air show performers or small teams.

2.4.2.2 Four Pillars of the Airshow-SMS

The *ASMS* retains the four fundamental components of the traditional SMS:

- **Safety Policy:** This is a performer's personal commitment to safety. It involves understanding and adhering to regulations and standards and developing procedures for handling emergencies.
- **Safety Risk Management:** Performers identify hazards associated with their routines, assess the associated risks, and implement methods to mitigate them. This could involve preflight checks, regular maintenance, rehearsal of routines, and contingency planning.
- **Safety Assurance:** Performers are encouraged to continuously monitor and evaluate their safety performance. This involves learning from any incidents or near misses and making necessary adjustments to routines or procedures.
- **Safety Promotion:** This involves seeking out and participating in relevant safety training, sharing safety information and experiences with other performers, and fostering a culture of safety within the community.

2.4.2.3 The Impact of an ASMS

Implementing an *ASMS* can bring about a significant enhancement in safety for individual performers and the wider air show community. This system provides a structured yet adaptable way to manage safety, helping performers identify and mitigate risks before they can lead to accidents or incidents.

2.4.2.4 Final Thoughts

The *ASMS* concept is an accessible and effective approach to safety management for individual performers and small teams within the air show community. By highlighting its adaptability and ease of implementation, we hope to encourage a stronger safety culture where all performers are actively managing risk, leading to safer, more successful performances.

As you continue reading, consider how the principles of the *ASMS* could be applied to your own routines and performances. Remember, a commitment to safety is the foundation of spectacular performances, and with the *ASMS*, we can ensure that these two elements go hand in hand.

The authors wish that this novel approach to SMS could enhance safety and reduce accidents for everyone involved in the air show industry, especially smaller organizations, and single-person air show performers.

2.4.3 The Insidious Hazard Associated with the Normalization of Deviance

Insidious hazard associated with a hazard is neither immediately evident nor conspicuous but instead progressively emerges or intensifies over time, frequently resulting in detrimental outcomes (Karachalios, 2022). Such hazards tend to be elusive, misleading, and challenging to identify, which contributes to their heightened danger. Insidious hazards can appear in various scenarios, such as finance, health, safety, or security, and often necessitate attentiveness, consciousness, and preemptive actions to reduce or avert potential damage.

Normalization of deviance, a concept initially proposed by American sociologist Diane Vaughan (2016), pertains to the incremental acceptance and assimilation of atypical conduct or practices typically deemed unsuitable or perilous. Normalization of deviance can be associated with insidious risk of high accident potential among all air show stakeholders, encompassing air show performers, air bosses, ACEs, event organizers, and regulators. Air shows require accuracy, rigor, and compliance with safety regulations. When deviations from SOPs and routine violations become normalized, the probability of accidents or incidents rises.

It has been suggested that the normalization of deviance played a critical role in the Challenger Space Shuttle disaster in 1986 (Vaughan, 2016). Engineers had identified an issue with the O-rings, which were not designed to withstand the low temperatures experienced during the launch. Despite this known issue, NASA and its contractors brushed aside the concerns of some engineers about the integrity of these rings when operated outside design temperature specifications.

They proceeded with the launch to meet the schedule resulting in the failure of the O-rings, leak of propellants, catastrophic explosion shortly after takeoff, and the loss of seven lives.

The normalization of deviance has also been identified as contributing to some mishaps within the air show industry, resulting in tragic outcomes for both operators and some spectators. These case studies exemplify what happens when the insidious risk associated with the normalization of deviance is not effectively managed:

2.4.3.1 Sknyliv Air Show Disaster (2002)

The Sknyliv Air Show disaster in Ukraine is an example of the normalization of deviance from the Su-27 pilots, which was never stopped by their supervisors. The pilots deviated from the planned routine and flew too low, resulting in a catastrophic crash that claimed at least 83 lives and injured nearly 600 spectators (BBC News, 2002). The normalization of performing maneuvers outside of established safety boundaries contributed to this tragedy.

2.4.3.2 Reno Air Races Crash (2011)

Another example is the Reno Air Race crash in Nevada, which resulted in 11 fatalities and at least 64 injured spectators. Pilot error and modifications to the aircraft, which deviated from established safety guidelines, contributed to the crash (National Transport Safety Board (NTSB), 2011). The normalization of these deviations ultimately led to the tragic accident.

2.4.3.3 Shoreham Airshow Crash (2015)

The Shoreham Airshow crash in the United Kingdom serves as another instance of normalizing deviance in the air show industry. The pilot attempted a loop maneuver at an altitude that was too low and insufficient for a safe recovery, causing a crash that left 11 fatalities and 13 injuries (Air Accidents Investigation Branch, 2017). The normalization of deviating from established safety guidelines and minimum altitudes contributed to this tragic accident.

The following recommendations can be helpful in reducing the normalization of deviance in air show operations:

- Encourage a proactive and positive safety culture: Organizations should prioritize safety and emphasize the importance of following established protocols. This includes fostering open communication and allowing team members to voice concerns about potential issues without fear of retribution.
- Affirm a culture of learning from past incidents: Analyzing previous accidents and incidents can help identify patterns of normalization of deviance and provide opportunities to implement corrective measures to prevent similar occurrences in the future.
- Implement an effective and generative reporting system: A robust reporting system should be in place to allow team members to report any deviations from established protocols or any potential risks they observe. This enables

organizations to identify trends and address issues before they escalate into accidents.

- Promote accountability and responsibility: Ensuring that everyone in the organization, from top management to ground crew, is accountable for maintaining safety standards can help mitigate the normalization of deviance. Responsibility should be clearly defined, and everyone should understand their role in upholding safety.
- Conduct periodic audits and inspections: External and internal audits or inspections should be carried out regularly to assess compliance with established safety protocols and identify areas for improvement. Experienced and impartial professionals should conduct these audits to ensure an unbiased evaluation of the organization's safety culture.

By recognizing all insidious risks and actively working to prevent the normalization of deviance, the air show industry can reduce the risk of accidents and ensure the safety of performers and spectators.

NOTES

1. A hammerhead turn, also known as a stall turn, is a maneuver performed by an aircraft in which the plane climbs vertically until it reaches the point of stall, and then falls back in the opposite direction while executing a 180-degree turn. The aircraft's nose points straight up during the climb, resembling the shape of a hammerhead, hence the name "hammerhead turn."

   The maneuver requires precise control and timing, as the aircraft must recover from the stall before it loses too much altitude.
2. Downline rolls are aerobatic maneuvers where the aircraft performs one or more rolls while flying along a downward trajectory. These maneuvers can be visually impressive but require precision and skill from the pilot. If not executed properly, downline rolls can lead to a significant loss of energy (low altitude and slow airspeed), loss of control, or incorrect positioning of the aircraft, potentially resulting in accidents.

LIST OF REFERENCES

Aero-News Network. (2013, June 24). *John Klatt down safe after engine loss during airshow.* Aero-News Network. http://www.aero-news.net/index.cfm?do=main.textpost&id=7f0e8522-e6c1-4aff-ba49-d3be9a18d1c7

Aero-News Network. (2022, November 7). *DHC-1 Chipmunk survives antenna collision during airshow.* http://www.aero-news.net/index.cfm?do=main.textpost&id=cf57feec-464a-486e-b610-78382da676a0

Air Accidents Investigation Branch. (2017). *Report on the accident to Hawker Hunter T7, G-BXFI near Shoreham airport on 22 August 2015* (Aircraft Accident Report No. 1/2017; p. 452). Department of Transport, United Kingdom. https://assets.publishing.service.gov.uk/media/58b9247740f0b67ec80000fc/AAR_1-2017_G-BXFI.pdf

Andersen, J. P., Papazoglou, K., Gustafsberg, H., Collins, P., & Arnetz, B. (2016, March 9). *Mental preparedness training.* FBI - The Law Enforcement Bulletin. https://leb.fbi.gov/articles/featured-articles/mental-preparedness-training

Athens Flying Week International Air Show (AFW). (2023). *Athens Flying Week International Air Show*. Athens Flying Week International Air Show. https://www.athensflyingweek.gr/

Aumeboonsuke, V., & Caplanova, A. (2021). An analysis of impact of personality traits and mindfulness on risk aversion of individual investors | SpringerLink. *Current Psychology*, *42*, 6800–6817.

Aviation Safety Network. (2007, September 1). *Accident Zlín 526AFS Akrobat Special SP-ELE*. Accident Zlín 526AFS Akrobat Special SP-ELE. https://aviation-safety.net/wikibase/21536

Aviation Safety Network. (2022, August 14). *Accident Hawker Hurricane Mk IV OO-HUR*. https://aviation-safety.net/wikibase/281539

Barker, D. (2003). *Zero error margin: Airshow display flying analysed*. https://www.researchgate.net/publication/261136368_Zero_Error_Margin_-_Airshow_Display_Flying_Analysed

Barker, D. (2020a). *Anatomy of airshow accidents* (1st ed., Vol. 1). International Council of Air Shows, International Air Show Safety Team.

Barker, D. (2020b, December). *Air show accidents a statistical and preventative perspective*. ICAS Convention. https://www.youtube.com/watch?v=29zT0edEAUQ&t=877s

BBC News. (2002, July 29). *Pilots blamed for Ukraine air disaster*. BBC News. http://news.bbc.co.uk/2/hi/europe/2159332.stm

BFU Switzerland. (2005). *Aircraft accident investigation report: Pilatus PC-21, HB-HZB, 13 January 2005* (Aircraft Accident Investigation Report No. 1909; p. 67). Aircraft Accident Investigation Bureau AAIB. https://www.sust.admin.ch/inhalte/AV-berichte/1909_d.pdf

Casey, J. (2007, December 13). *Fatal 2006 air show crash*. Oregonlive.com. https://www.oregonlive.com/washingtoncounty/2007/12/fatal_2006_air_show_crash.html

Chen, C.-F., & Chen, S.-C. (2014). Measuring the effects of Safety Management System practices, morality leadership and self-efficacy on pilots' safety behaviors: Safety motivation as a mediator. *Safety Science*, *62*, 376–385. https://doi.org/10.1016/j.ssci.2013.09.013

Chionis, D., & Karanikas, N. (2018). Differences in risk perception factors and behaviours amongst and within professionals and trainees in the aviation engineering domain. *Aerospace*, *5*, Article 2. https://doi.org/10.3390/aerospace5020062

Crundall, D., van Loon, E., Stedmon, A. W., & Crundall, E. (2013). Motorcycling experience and hazard perception. *Accident Analysis & Prevention*, *50*, 456–464. https://doi.org/10.1016/j.aap.2012.05.021

Daily Sabah. (2022, December 6). *Military plane crashes in central Türkiye, pilot survives*. Daily Sabah. https://www.dailysabah.com/turkey/military-plane-crashes-in-central-turkiye-pilot-survives/news

Dekker, S. (2014). *Safety differently: Human factors for a new era* (2nd ed.). CRC Press.

Dolbeer, R. A. (2011). Increasing trend of damaging bird strikes with aircraft outside the airport boundary: Implications for mitigation measures. *Human–Wildlife Interactions*, *5*(2), 235–248.

Drinkwater, J., & Molesworth, B. (2010). Pilot see, pilot do: Examining the predictors of pilots' risk management behaviour. *Safety Science*, *48*(10), 1445–1451. https://doi.org/10.1016/j.ssci.2010.07.001

Efthymiou, M., Whiston, S., O'Connell, J. F., & Brown, G. D. (2021). Flight crew evaluation of the flight time limitations regulation. *Case Studies on Transport Policy*, *9*(1), 280–290.

Federal Aviation Administration. (2020a). Chapter 6: Issue a certificate of waiver or authorization for an aviation event. In *Flight standards information management system* (pp. 3-141-3–160). Federal Aviation Administration. https://fsims.faa.gov/wdocs/8900.1/v03%20tech%20admin/chapter%2006/03_006_001.htm

Federal Aviation Administration. (2020b). *Order 8000.369C – Safety Management System*. Federal Aviation Administration (FAA). https://www.faa.gov/regulations_policies/orders_notices/index.cfm/go/document.current/documentNumber/8000.369

Federal Aviation Administration. (2021). *Airplane Flying Handbook (FAA-H-8083-3C)*. Federal Aviation Administration (FAA). https://www.faa.gov/regulations_policies/handbooks_manuals/aviation/airplane_handbook

Federal Aviation Administration. (2022). *Risk Management Handbook (FAA-H-8083-2A)*. Federal Aviation Administration (FAA). https://www.faa.gov/regulationspolicies/handbooksmanuals/risk-management-handbook-faa-h-8083-2a

Ferraro, R., VanDyke, D., Zander, M., Anderson, K., & Kuehlen, B. (2015). Risk perception in aviation students: Weather matters. *International Journal of Aviation,Aeronautics, and Aerospace*, *2*(1). https://doi.org/10.15394/ijaaa.2015.1044

Gibson, R. J., Michayluk, D., & Van de Venter, G. (2013). Financial risk tolerance: An analysis of unexplored factors. *Financial Services Review*, *22*, 23–50. https://opus.lib.uts.edu.au/handle/10453/23532

Hallahan, T. A., Faff, R. W., & McKenzie, M. D. (2004). An empirical investigation of personal financial risk tolerance. *Financial Services Review*, *13*, 57–78. https://doi.org/10.1.1.392.58

Harris, D. (2011). *Human performance in the flight deck*. CRC Press.

Heath, C., & Heath, D. (2013). *Decisive: How to make better choices in life and work*. Crown Publishers.

Hollnagel, E. (2006). *Resilience: The challenge of the unstable*. Ashgate. http://urn.kb.se/resolve?urn=urn:nbn:se:liu:diva-38166

Hunter, D. R. (2002). *Risk perception and risk tolerance in aircraft pilots* (Technical Report DOT/FAA/AM-02/17; p. 26). Office of Aerospace Medicine, Federal Aviation Administration. https://www.researchgate.net/profile/David-Hunter-27/publication/235148351_Risk_Perception_and_Risk_Tolerance_in_Aircraft_Pilots/links/00b7d518ab8b52526a000000/Risk-Perception-and-Risk-Tolerance-in-Aircraft-Pilots.pdf

ICAO. (2018). *Doc 9859, safety management manual*. International Civil Aviation Organization (ICAO). https://skybrary.aero/sites/default/files/bookshelf/5863.pdf

International Council of Air Shows. (n.d.). *International Council of Air Shows*. International Council of Air Shows (ICAS). Retrieved May 11, 2023, from https://airshows.aero/

International Council of Air Shows. (2012, March 8). *ICAS individual risk management worksheet*. ICAS. https://airshows.aero/ViewDoc/2781

International Council of Air Shows. (2023). *Air Boss Recognition Program (ABRP) Manual*. International Council of Air Shows. https://airshows.aero/GetDoc/4140

Jensen, R. S., & Benel, R. A. (1977). *Judgment evaluation and instruction in civil pilot training* (No. FAA-RD-78-24 Final Rpt.). https://trid.trb.org/view/75609

Ji, M., You, X., Lan, J., & Yang, S. (2011). The impact of risk tolerance, risk perception and hazardous attitude on safety operation among airline pilots in China. *Safety Science*, *49*(10), 1412–1420. https://doi.org/10.1016/j.ssci.2011.06.007

Joseph, C., & Reddy, S. (2013). Risk perception and safety attitudes in Indian army aviators. *The International Journal of Aviation Psychology*, *23*(1), 49–62. https://doi.org/10.1080/10508414.2013.746531

Kaplan, R. S., & Mikes, A. (2012, June). Managing risks: A new framework. *Harvard Business Review*. https://www.hbs.edu/faculty/Pages/item.aspx?num=42549

Karachalios, E. "Manolis." (2022). *An evaluation of the relationships between resilient safety culture, safety risk parameters, and mindfulness in the international air show community* [PhD Dissertation, University North Dakota]. https://commons.und.edu/theses/4267/

Lam, J. (2014). *Enterprise risk management: From incentives to controls* (2nd ed.). Wiley. https://www.wiley.com/en-us/Enterprise+Risk+Management%3A+From+Incentives+to+Controls%2C+2nd+Edition-p-9781118413616

Martinussen, M., & Hunter, D. R. (2018). *Aviation psychology and human factors* (2nd ed.). CRC Press.

Metzler, M. M. (2020). G-LOC due to the push-pull effect in a fatal F-16 mishap. *Aerospace Medicine and Human Performance, 91*(1), 51–55. https://doi.org/10.3357/AMHP.5461.2020

National Transport Safety Board. (2022). *Mid-air collision between a B-17 and P-63* (Aircraft Accident Investigation Report CEN23MA034; p. Not available). National Transport Safety Board. https://www.ntsb.gov/investigations/Pages/CEN23MA034.aspx

National Transport Safety Board (NTSB). (2011). *Aviation accident at the Reno Air Race* (Aircraft Accident Investigation Report WPR11MA454; p. 52). National Transport Safety Board. https://www.ntsb.gov/investigations/pages/WPR11MA454.aspx

Newman, D. G. (1997). +GZ-induced neck injuries in Royal Australian Air Force fighter pilots—PubMed. *Aviat Space Environ Med, 68*(6), 520–524.

O'Hare, D. (1990). Pilots' perception of risks and hazards in general aviation. *Aviation, Space, and Environmental Medicine, 61*(7), 599–603.

Radio Free Europe/ Radio Liberty. (2022, July 27). *Remembering Skynliv: The deadliest air show disaster in history.* Remembering Skynliv: The Deadliest Air Show Disaster in History. https://www.rferl.org/a/remembering-skynliv-deadliest-air-show-aviation-history/31961756.html

Reason, J. (1991). *Human error* (1st ed.). Cambridge University Press.

Royal Air Force, United Kingdom. (2023). *Red Arrows, Royal Air Force.* Red Arrows, Royal Air Force. https://www.raf.mod.uk/display-teams/red-arrows/

Royal Canadian Air Force. (2019, December 6). *Canadian Forces Snowbirds return to operations* [Not available]. Royal Canadian Air Force. https://www.canada.ca/en/department-national-defence/maple-leaf/rcaf/2019/12/canadian-forces-snowbirds-return-to-operations.html

Scheid, T. L., & Wright, E. R. (Eds.). (2017). *A handbook for the study of mental health* (3rd ed.). Cambridge University Press. https://www.cambridge.org/gr/academic/subjects/psychology/health-and-clinical-psychology/handbook-study-mental-health-social-contexts-theories-and-systems-3rd-edition?format=HB&isbn=9781107134874

Schopf, A. K., Stouten, J., & Schaufeli, W. B. (2021). The role of leadership in air traffic safety employees' safety behavior. *Safety Science, 135*, 105118. https://doi.org/10.1016/j.ssci.2020.105118

Sjoberg, L., Moen, B.-E., & Rundmo, T. (2004). *Explaining risk perception. An evaluation of the psychometric paradigm in risk perception research* (p. 39). Norwegian University of Science adn Technology, Department of Psychology.

Stolzer, A. J., & Goglia, J. J. (2016). *Safety management systems in aviation* (2nd ed.). Routledge.

Tchankova, L. (2002). Risk identification – Basic stage in risk management. *Environmental Management and Health, 13*(3), 290–297. https://doi.org/10.1108/09566160210431088

UK Air Accidents Investigation Branch. (2017). *Aircraft accident report AAR 1/2017—G-BXFI, 22 August 2015* (Aircraft Accident Report AAR 1/2017). UK Air Accidents Investigation Branch. https://www.gov.uk/aaib-reports/aircraft-accident-report-aar-1-2017-g-bxfi-22-august-2015

UK Civil Aviation Authority. (2023a). *CAP 403: Flying displays and special events: Safety and administrative requirements and guidance*. UK Civil Aviation Authority. https://publicapps.caa.co.uk/docs/33/CAP403%20Edition%2020.pdf

UK Civil Aviation Authority. (2023b). *Introduction to bowtie: How this risk assessmeent tool works*. Introduction to Bowtie. https://www.caa.co.uk/safety-initiatives-and-resources/working-with-industry/bowtie/about-bowtie/introduction-to-bowtie/

U.S. Air Force. (1994). *Summary of AFR 110-14 USAF accident investigation board report* (Aircraft Accident Investigation Report No. 110–14). United States Air Force. https://www.baaa-acro.com/crash/crash-boeing-b-52h-stratofortress-fairchild-afb-4-killed

U.S. Air Force. (2018). *United States Air Force aircraft accident investigation board report, F-16CM, T/N 91-0413, 4 April 2018* (p. 37) [Aircraft Accident Investigation Report]. United States Air Force. https://www.acc.af.mil/Portals/92/AIB/180404_ACC_Creech_F16_Thunderbird_AIB_Narrative_Report.pdf?ver=2018-10-16-132501-787

U.S. Navy. (2016). *Report on June fatal Blue Angels crash*. U.S. Naval Institute. https://news.usni.org/2016/09/16/document-report-june-fatal-blue-angels-crash

Vaughan, D. (2016). *The challenger launch decision: Risky technology, culture, and deviance at NASA*. University of Chicago Press. https://press.uchicago.edu/ucp/books/book/chicago/C/bo22781921.html

Wagstaff, P. (2016, February 6). *The life of an air show pilot*. Plane & Pilot Magazine. https://www.planeandpilotmag.com/article/the-life-of-an-air-show-pilot-2/

Wiegmann, D. A., & Shapell, S. A. (2016). *A human error approach to aviation accident analysis: The human factors analysis and classification system* (1st ed.). Routledge. https://www.taylorfrancis.com/books/mono/10.4324/9781315263878/human-error-approach-aviation-accident-analysis-douglas-wiegmann-scott-shappell

Winter, S., Truong, D., & Keebler, J. (2019). The flight risk perception scale (FRPS): A modified risk perception scale for measuring risk of pilots in aviation. *International Journal of Aviation Research*, *11*(1), 57–72. https://doi.org/10.22488/okstate.19.100407

You, X., Ji, M., & Han, H. (2013). The effects of risk perception and flight experience on airline pilots' locus of control with regard to safety operation behaviors. *Accident; Analysis and Prevention*, *57C*, 131–139. https://doi.org/10.1016/j.aap.2013.03.036

3 Air Show Culture

3.1 INTRODUCTION

One might wonder about the relevance of including a chapter on air show culture in this book. Nevertheless, we find it essential to familiarize the reader with the distinct characteristics of this community, which distinguishes it from other constituents within the aviation industry.

Similar to national and professional cultures, the air show culture exhibits variations compared to other sectors in the aviation field. Although there seem to be no empirical studies to substantiate these differences, one of the authors has personal experiences as an active member of the air show community for over a decade, which provides valuable inputs to identify specific values, beliefs, and practices unique to this domain.

The air show culture is exclusive when compared to other organizations, primarily due to the following reasons:

- Passion for Aviation

The air show community is bound together by their shared love for aviation, which is the cornerstone of their activities. This passion unites individuals with diverse backgrounds and skill sets who come together to showcase aerobatics and highlight technological innovations in the aviation industry.

- Sense of Family

The air show culture fosters a unique sense of camaraderie and family, where members support and encourage each other. This atmosphere is nurtured by shared experiences, memories, and a mutual commitment to safety, professionalism, and teamwork.

- High-Risk Environment

Air shows involve high-risk performances, requiring an extraordinary level of skill, professionalism, and safety measures. This aspect creates a strong sense of responsibility and interdependence among community members, who work together to ensure the success and safety of each event.

- Public Engagement

Air show performers and organizers prioritize engaging with the public and sharing their passion for aviation. They actively work to inspire the next generation of pilots

DOI: 10.1201/9781003431879-3

and aviation enthusiasts, making their events accessible and enjoyable for a broad audience.

- Rich History and Tradition

Air shows have a long-standing history and tradition, with many community members remaining involved for decades. This continuity strengthens the bonds between participants and deepens the sense of family within the community.

While other organizations may share some of these characteristics, the combination of all these factors makes the air show culture distinctly unique and special.

Furthermore, an integral part of the air show culture is a culture of safety and excellence. By providing an overview of the unique aspects of the air show culture, this chapter aims to offer readers valuable insight into the remarkable characteristics that set it apart from other fields within the aviation industry. During the COVID-19 pandemic, resilient safety culture, which is an essential component of the overall air show culture, was assessed within the air show industry. In this chapter, the research findings focus on resilient safety culture and its relationships with safety risk factors, and mindfulness is discussed.

3.2 SAFETY CULTURE

> The term "safety culture" has been widely applied in numerous industries and organizations, including aviation, yet there is no consistent definition for safety culture in the literature. *Safety culture* is described by the United States Nuclear Regulatory Commission (2020) as the core values and behaviors that result from a mutual effort by leaders and individuals to prioritize safety over competing interests to protect people and the environment, while as per the United Kingdom Health and Safety Executive (2002), safety culture encompasses an organization's behavioral and situational aspects. Moreover, safety culture reflects a group's shared values, customs, assumptions, and outlooks related to safety and risk (Federal Aviation Administration, 2022; Mearns & Flin, 2018; Yorio et al., 2019).

Extant research (Akselsson et al., 2009a; Clarke, 2000; Glendon & Stanton, 2000; Hofstede et al., 2010; Hollnagel, 2014; Merritt & Helmreich, 1996; Patankar et al., 2012) suggested that safety culture is a subculture of other more significant cultures for an organization, similar to an onion with many layers of skin. National culture or another primary culture could be the outer layer, while the employee is in the center. At times, conflicting cultural values might exist (Liao, 2015); therefore, any safety culture model should extend beyond the organization's boundaries (Harris, 2011, p. 284).

Reason (1997) suggested that few concepts have been studied so much and yet understood so little as the safety culture. If organization members are convinced that they have an adequate level of safety culture, they are usually mistaken; for Reason, safety culture involves continuous effort and is rarely achieved. Yet, a strong safety culture can help improve the safety performance of an organization (Shirali et al., 2016).

The UK CAA (2023) considers a positive safety culture as crucial to a safe flying display community within the air show community. This culture is influenced by a number of factors, including the behaviors of authorities, flying display directors, and display authorization examiners; the adherence of display pilots to established standards; and the encouragement of open and honest reporting of any incidents that may result in the transmission of lessons learned. Moreover, a tangible performance metric of safety management systems (SMS) implementation is the fostering of a positive safety culture throughout the organization, characterized among other attributes by safety reporting and information sharing (Federal Aviation Administration, 2022).

Building a safety culture is a process that needs gradual and collective efforts from all parties in an organization (Federal Aviation Administration, 2022; Hollnagel, 2014). Reason (1997) suggested that safety culture is a multifaceted entity consisting of the following interacting elements: a reporting culture, a just culture, a flexible culture, and a learning culture. These elements interact to create an informed culture, which is the basis for the term "safety culture." Individuals at all levels in an organization should recognize their safety responsibilities and must be accountable for their actions to promote a safety culture (Federal Aviation Administration, 2022; ICAO, 2018).

Nevertheless, Akselsson et al. (2009) had earlier suggested that the concept of safety culture required more context in discussions to address some identified limitations in scope. The reason for this could be the apparent focus by some organizations on singular aspects of such a multidimensional construct (an example is the focus on just culture as representative of the entire safety culture dimension). Moreover, organizations advocating for a strong safety culture might disregard the aspect of resilience due to a lack of management commitment or communication. Finally, inconsistencies between what is stated or written and what is practiced might result in gaps in an organization's safety culture (Adjekum & Fernandez Tous, 2020).

In addition to these suggested limitations, Shirali et al. (2016) suggested that there is minimal to no standardized approach to deriving qualitative attributes to describe the safety culture dimension. The existing dimensions used to measure safety culture are based on concepts such as behavior, values, assumptions, and norms while excluding any dynamic interactions among people, technology, and administration.

3.3 RESILIENT SAFETY CULTURE

The concept of resilient safety culture has been suggested to provide better context in the discussion on safety culture (Weick & Sutcliffe, 2001, 2009). Shirali et al. (2016) defined *resilient safety culture* as a safety culture that focuses on resilience, learning, continuous enhancements, and cost-effectiveness. According to Akselsson et al. (2009), resilience safety is no different in theory from a safety culture, and the main difference lies in how it is practised. Akselsson et al. (2009b, p. 4) also provided a more thorough definition of *resilience safety culture*: "Resilience safety culture is an organizational culture that fosters safe practices for improved safety in an ultra-safe organization striving for cost-effective safety management by stressing the resilience engineering, organizational learning, and continuous improvements."

Thus, resilient safety culture has some characteristics that differentiate it from a safety culture in an organization. According to Shirali et al. (2015), the attributes of resilient safety culture in an organization are situational adaptability, institutional learning, continuous improvements, and cost-effectiveness in operations. Moreover, resilient safety culture is based on three types of capabilities: psychological/cognitive, behavioral, and managerial/contextual (Pillay et al., 2010).

Akselsson et al. (2009) stated that an organization with a resilient safety culture has the following characteristics:

- It emphasizes the need for a learning culture backed by a just culture.
- It strives for resilience and develops and uses forward feed control to keep processes within safe limits.
- It strives for efficiency in safety management and integrates safety and core business performance.

Additionally, a culture of reliability may function in concert with certain fundamental organizational traits, allowing mindfulness, a distinguishing attribute of high-reliability organizations (HROs), to thrive and grow (Cantu et al., 2020).

The enhancement of a strong and proactive resilient safety culture can help not only to improve the safety performance of an organization but also to recover from an upset (Shirali et al., 2016). Even when incidents do occur, a resilient safety culture can enable an organization to adapt, successfully recover, and operate effectively within the margins of safety (Hollnagel, 2014).

The concept of resilient safety culture in the aviation industry has been the focus of several studies (Adjekum & Fernandez Tous, 2020; Akselsson et al., 2009a; Heese et al., 2014; Hollnagel, 2014; Hollnagel et al., 2011; Reason, 2016; Teske & Adjekum, 2022). In general, these studies advocate for a resilient safety culture as essential in improving an organizational culture that fosters safe practices.

The concept of resilient safety culture has also been proposed in other industries, such as the health care and petrochemical industries. Smith and Arfanis (2013) have suggested that the safe delivery of health care is fundamentally derived from cultivating individual and organizational resilient safety cultures. Sujan et al. (2019) posit that a resilient safety culture allows healthcare systems to continually adapt to changing circumstances. In the petrochemical industry, Shirali et al. (2016) concluded that adopting opportunities in safety culture and resilience engineering can drive improvements in safety performance.

3.4 AIR SHOW PERFORMERS' EXCELLENCE CULTURE

A key component of air show culture is an unwavering commitment to excellence, which is deeply ingrained in the community's core values. This dedication to excellence encompasses not only safety but also the pursuit of precise performances and exceptional results across all aerial events. All in all, the air show community's existing values, experiences, training, and accomplishments come together to create an environment that continually strives for perfection.

3.4.1 Defining Excellence Culture for Air Show Performers

Excellence culture, in the context of air show performers, refers to the values, beliefs, and behaviors that promote exceptional performance, continuous improvement, and collaboration among team members (Schein & Schein, 2016). This culture is characterized by a strong commitment to achieving excellence in all aspects of air show performances, from pilot skills and aircraft maintenance to display profile "choreography" and audience engagement.

3.4.2 Key Elements of Air Show Excellence Culture

The air show industry is renowned for its thrilling aerial displays that captivate audiences worldwide. Behind the scenes, a culture of excellence is necessary for air show performers to consistently deliver high-quality performances while ensuring safety and professionalism. This section explores the key elements of an air show excellence culture, such as expertise and skill development, teamwork and collaboration, and audience engagement and innovation.

Expertise and skill development are fundamental attributes of an excellent culture for air show performers. Continuous improvement of piloting skills, aircraft maintenance knowledge, and performance techniques is crucial for them to excel. For example, renowned teams such as the USAF Thunderbirds, UK RAF Red Arrows, Royal Canadian Air Force Snowbirds, and the US Navy Blue Angels continually refine their skills through rigorous training and practice sessions. These elite performers invest time and effort into perfecting their craft, ensuring that they can execute complex maneuvers with precision and confidence.

Teamwork and collaboration form the backbone of a culture of excellence. Air show performers rely on seamless cooperation among pilots, ground crew, and event organizers to execute their routines flawlessly. This collaboration ensures the smooth execution of complex maneuvers, such as the Red Arrows' signature Diamond Nine formation (see Figure 3.1). Furthermore, strong communication and trust among team members are vital to the success and safety of each performance.

Safety risk management is an essential component of an air show culture of excellence. Given the inherent risks involved in aerial displays, prioritizing safety and implementing effective risk management strategies are crucial. Organizations like the International Council of Air Shows (ICAS) play a significant role in promoting safety standards and best practices within the air show industry. Through their

FIGURE 3.1 Red Arrows in the Diamond 9 formation at RIAT 2013. Copyright: Tim Felce.

efforts, air show performers are equipped with the knowledge and tools to minimize risks and ensure a safe environment for themselves and the audience.

Audience engagement and innovation are critical factors that contribute to a culture of excellence for air show performers. By creating memorable experiences for audiences through engaging performances and innovative aerial displays, performers can showcase their talents and solidify their reputations within the industry. Teams like the Patriot Jet Team captivate audiences with their precise formation flying and unique aircraft designs. Incorporating new and creative elements into their routines helps performers maintain a sense of excitement and wonder for their audiences, further reinforcing the culture of excellence that drives their success.

In conclusion, a culture of excellence is paramount for air show performers as it encompasses the essential elements required for their success and safety. Expertise and skill development, teamwork and collaboration, safety and risk management, and audience engagement and innovation all contribute to the creation and maintenance of this culture. By embracing these elements, air show performers can continue to captivate audiences worldwide while ensuring their performances are of the highest quality and safety standards.

3.4.3 Fostering a Culture of Excellence among Air Show Performers: Strategies

The pursuit of excellence is an essential aspect of the air show industry, as it directly impacts safety, performance, and overall success. To foster a culture of excellence among air show performers, several strategies must be implemented, including establishing clear goals and objectives, encouraging continuous learning and skill development, promoting effective communication and feedback, and recognizing and celebrating excellence.

3.4.3.1 Establishing Clear Goals and Objectives

The first step toward fostering a culture of excellence among air show performers is to establish clear goals and objectives for their performances. These goals should be specific, measurable, achievable, relevant, and time-bound (SMART), allowing performers to align their actions and behaviors with the pursuit of excellence. A typical practical goal for an air show performer might be: "To successfully execute a new and thrilling aerial maneuver at the upcoming air show, engaging and entertaining the audience while maintaining safety standards." Using the Ribbon Cut[1] maneuver as the aerial maneuver, a SMART goal for an air show performer could be: "To successfully perform the Ribbon Cut maneuver at the upcoming air show in three months, cutting the ribbon at a 20-foot altitude, while maintaining safety standards and captivating the audience."

Applying the SMART criteria to this goal:

- Specific: The goal is to perform the Ribbon Cut maneuver at the upcoming air show in three months, cutting the ribbon at a 20-foot altitude. This gives the performer a precise focus on what they need to achieve.
- Measurable: The goal can be measured by the successful completion of the Ribbon Cut maneuver during the air show performance, cutting the ribbon at the target altitude, as well as by audience engagement and feedback.
- Achievable: The performer should have the necessary skills and experience to perform the Ribbon Cut maneuver. They should also ensure they have enough time and resources to practice and perfect the maneuver before the air show.
- Relevant: The goal is directly related to the performer's role as an air show performer, and it contributes to their overall objective of providing an engaging and entertaining show for the audience.
- Time-bound: The goal must be accomplished by the date of the upcoming air show in three months. This provides a clear deadline for the performer to work toward, helping them stay focused, and motivated throughout the preparation process.

Clearly defined objectives serve as a road map for performers, helping them understand the expectations and desired outcomes of their performances. This clarity also allows performers to focus their efforts on achieving these objectives, resulting in higher levels of performance and a greater sense of accomplishment.

3.4.3.2 Encouraging Continuous Learning and Skill Development

Continuous learning and skill development are crucial to maintaining a culture of excellence among air show performers. By providing resources and opportunities for pilots and crew members to enhance their skills and knowledge, air show organizations can ensure that their performers are well-prepared and continually improving.

This can include offering training programs, workshops, seminars, and access to cutting-edge technology and tools, which may consist of advanced flight simulators and virtual and augmented reality (VR/AR) trainers, which can help performers refine their abilities and adapt to changes in the industry.

3.4.3.3 Promoting Effective Communication and Feedback

Open and transparent communication channels are essential for fostering a culture of excellence. Effective communication and feedback systems among team members promote trust, collaboration, and a shared understanding of safety and performance standards. This can be achieved through regular team meetings, debriefs after performances, and the use of communication technologies that enable real-time information sharing, such as social media platforms. By encouraging open dialogue, performers can share their experiences, learn from each other, and continuously refine their skills and performance.

3.4.3.4 Recognizing and Celebrating Excellence

Acknowledging and rewarding air show performers who demonstrate exceptional performance and contribute to their team's success reinforces a culture of excellence. This can be achieved through various means, such as formal recognition programs, awards, or informal gestures of appreciation. Celebrating excellence not only motivates performers to continue striving for improvement but also serves as a reminder of the high standards expected within the air show community.

In conclusion, fostering a culture of excellence among air show performers is vital for ensuring safety, outstanding performance, and overall success in the industry. By implementing strategies such as establishing clear goals and objectives, encouraging continuous learning and skill development, promoting effective communication and feedback, and recognizing and celebrating excellence, air show organizations can create an environment where performers are motivated to pursue and maintain excellence in all aspects of their work.

3.4.3.5 Examples of Excellence Culture in Air Show Performers

The examples below demonstrate how a culture of excellence among air show performers can lead to outstanding achievements, enhanced audience engagement, and a strong focus on safety. By adopting the strategies outlined in this report, performers can work toward cultivating an excellent culture that benefits the entire air show community.

- Aude Lemordant: A world-champion aerobatic pilot, Aude Lemordant, has inspired countless aviation enthusiasts through her remarkable air show performances. She is known for her dedication to perfecting her craft, pushing the boundaries of aerobatic flying, and promoting safety within the air show community. As a trailblazer in the field, she continues to break barriers and raise the bar for excellence.
- The Flying Bulls: Owned by Red Bull, The Flying Bulls is a group of experienced pilots who perform stunning aerial displays in various air shows

worldwide. Although the Red Bull Air Race World Championship is no longer active, The Flying Bulls continue to demonstrate a commitment to excellence, showcasing their exceptional flying skills, precision, and adherence to safety standards.

- The Frecce Tricolori: As the aerobatic demonstration team of the Italian Air Force, the Frecce Tricolori represents a culture of excellence through their highly skilled pilots, intricate formation flying, and dedication to safety. They continually strive to improve and innovate their performances, earning them international recognition and acclaim (see Figure 3.2).

3.4.3.6 Excellence and Safety: A Continuous Juggling Act of Balance

Excellence is the driving force for air show performers, who consistently aim for high standards in performance, experience, and outcomes. In this context, excellence goes beyond merely adhering to safety guidelines; it sets the stage for showcasing remarkable skill, precision, and artistry in each performance.

Fostering a culture of excellence serves as the cornerstone for ensuring safety and superior quality throughout air show events. The need for excellence in air shows serves as an impetus to uphold high-performance standards that surpass average expectations and minimum regulatory safety requirements. This commitment extends to all aspects of air shows, including planning, communication, and coordination among organizers, pilots, and support teams.

In the realm of air shows, excellence and safety work together in harmony, reinforcing each other to ensure the highest levels of performance and audience satisfaction. This symbiotic relationship between excellence and safety not only minimizes risks but also elevates the overall quality and impact of air shows. By focusing on excellence, air shows become memorable experiences that delight everyone involved.

Ultimately, an excellent air show performance, characterized by precise execution and a skilled showcase of the full capabilities of an aircraft, can inherently guarantee a safe experience. However, the inverse is not necessarily true; a safe air show doesn't

FIGURE 3.2 The Frecce Tricolori showing the Italian Flag over the Volkel Airshow 2013. Photo by the author (Manolis Karachalios).

automatically equate to excellence. In some cases, a safe air show might involve a performer doing the absolute minimum to demonstrate their aircraft's capabilities, which could result in an underwhelming or lackluster performance.

3.5 PERSPECTIVES FROM AIR SHOW PERFORMERS ON AIR SHOW CULTURE

3.5.1 Interviews Results

To understand the perspectives of air show performers on the air show industry culture, relevant findings from interviews and focus group sessions of the 2020 air show performers study are discussed in this section. The themes identified were operational culture, ownership, and continuous enhancements (see Figure 3.3).

3.5.1.1 Operational Culture

The operational culture theme is undergirded by notable codes such as safety culture, excellence culture, disciplined culture, and culture of change in air show operations. The majority of air show performers described the existing safety culture in their air show operations as robust, with significant improvements over time. Some air show performers intimated that different countries have developed safety management systems (SMS) and programs based on a recommendation from ICAO standards, but the approach to maintaining a proactive safety culture may vary (see Figure 3.3).

3.5.1.1.1 Change Culture

Historically, the air show community had a pervasive disposition toward hazardous attitudes such as machismo, invulnerability, and antiauthority. These hazardous attitudes led to at-risk and sometimes outright reckless behaviors during flight displays. Some of these displays appealed to spectators whose encouragement egged on such performances leading to safety events and, in some cases, fatalities to both performers and spectators. However, in the period after the Shoreham air disaster in 2015, there seems to have been some positive changes in both personal and organizational culture within the air show community. The changes in culture have led to a shift in mentoring new air show performers, with a greater focus on safety and learning from experienced pilots. This change demonstrates the industry's ongoing cultural transformation toward safety and excellence.

3.5.1.1.2 Safety Culture

The discussion on the safety culture themes by air show performers was mostly related to three aspects of a safety culture based on Reason's (1997) framework: reporting culture, learning culture, and informed culture. The majority of air show performers stated that the current reporting culture in the air show community relies on peer-to-peer interactions due to its small and tight-knit nature. This informal reporting system allows for quick resolution of safety concerns, but formal reporting systems vary by country.

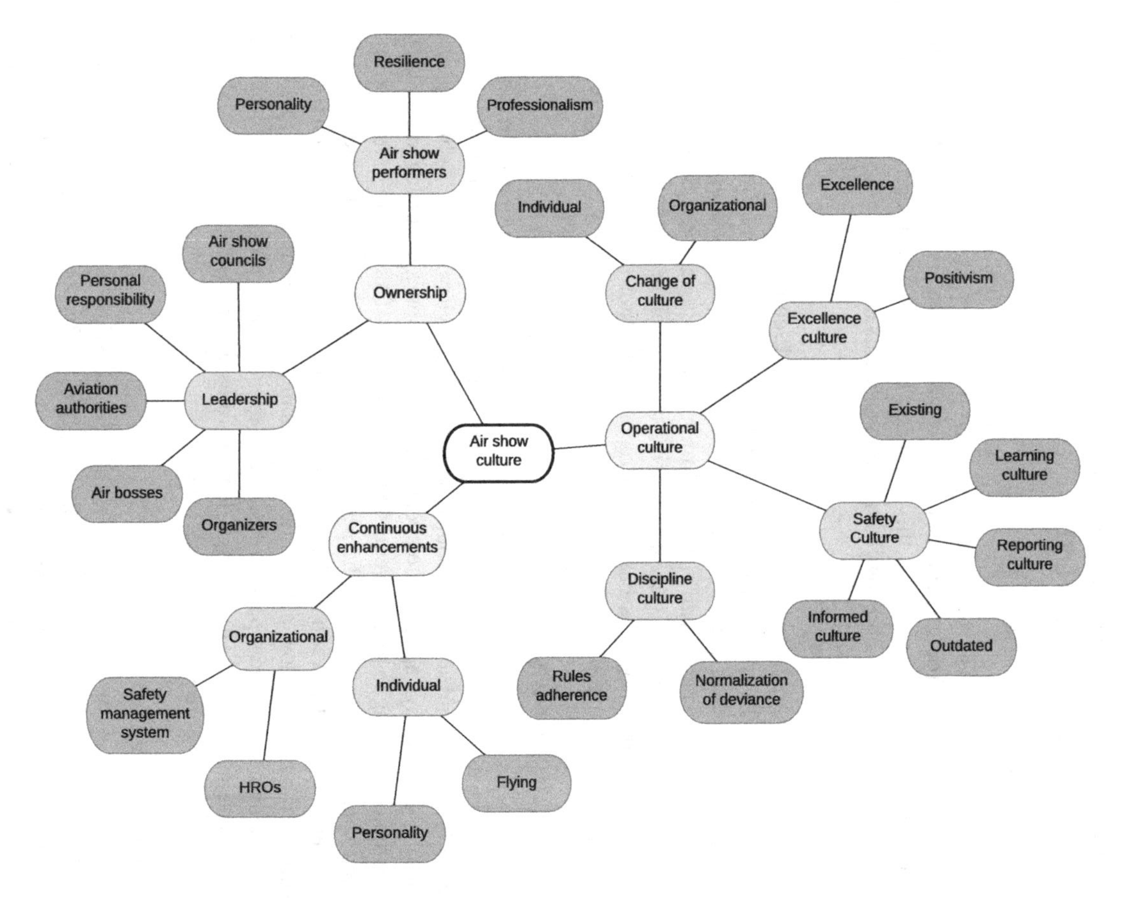

FIGURE 3.3 Air show culture, theme map.

One air show performer from Africa mentioned the lack of a standardized safety system with formal documentation, while another respondent discussed the advantages and disadvantages of having an anonymous reporting system. Implementing such an anonymous reporting system would be challenging due to the diverse backgrounds, activities, aircraft types, and teams involved in air shows. Although possible, introducing more paperwork for display pilots might not be the most effective solution, as there is already a considerable workload for them to manage.

3.5.1.1.3 Transmissivity of Experience

Another air show performer with extensive experience as both a pilot and a flying display director emphasized the importance of preserving institutional memory and transferring experience within the air show community. This individual suggested that one's legacy could include promoting the exchange of experiences and ideas to ensure the continuation of learning and growth in the field.

Many respondents highlighted the significance of instructors, not just during initial training but also as mentors throughout a display pilot's career. For example, some respondents mentioned receiving invaluable advice from their mentor, which helped them develop as professional air show pilots. Another respondent suggested making educational resources available online for those interested in becoming air show pilots.

Regarding informed culture, two experienced air show performers agreed on the vital role of communication and information exchange among all stakeholders in the air show community. One respondent stressed the importance of experienced pilots offering guidance to less experienced pilots, while another emphasized communication and education as crucial components for improving safety culture. The same respondent also suggested that interactive workshops at conventions could be more effective than presentations, as they provide an opportunity for stakeholders, including regulators, organizers, performers, and sponsors, to discuss problems and challenges openly. Additionally, a respondent undergoing training to become a fast jet display pilot expressed appreciation for safety workshops where air show performers can discuss safety issues and share their experiences.

3.5.1.1.4 Culture of Excellence

Air show performers mentioned the presence of a culture of excellence within the current air show community, demonstrated by high standards among operators. Some respondents shared their experience of working with top-caliber individuals, organizers, and team members in which excellence was a desired attribute. Another air show performer stated that it is not only the air show performers who have changed their minds about operational risk and the culture of excellence but also display evaluators and Aerobatic Competency Evaluators (ACE) who have embraced this culture of excellence and have raised the authorization standards by increasing the stringency requirements when giving display authorizations to pilots to protect not only the pilots or the air show industry but also to save the lives of those who enjoy watching air shows.

Some respondents stressed the value of sharing best practices related to air show displays among performers, organizers, and air bosses. One respondent suggested more focus on the positive aspects of air show performance and needed improvements rather than just always pointing out negatives, as this may yield a better behavioral change.

Another respondent agreed with the previous assertion, expressing their preference for positive reinforcement that emphasizes good practices. An experienced air show pilot highlighted the importance of shifting the mentality from a safety culture to a culture of excellence, which can be reflected in various aspects of the air show community, such as patches on flight suits symbolizing a commitment to excellence.

3.5.1.1.5 Discipline Culture

Many respondents highlighted the importance of fostering professionalism in the air show community, with a particular focus on enhancing operational discipline while addressing the issue of normalization of deviance and its potential impact on flight safety. One respondent emphasized the need for a culture of discipline that does not tolerate deviations from established rules and regulations for air show displays. The respondent advocated for deviations to be immediately addressed and rectified.

Another respondent intimated that the consequences of allowing the normalization of deviance to persist are the likelihood of accidents and loss of life at air shows. An experienced air show performer stressed the importance of adhering to rules and implementing a zero-tolerance policy, not just for performers but also for air bosses. One interviewee with vast experience as an air show performer, air boss, and part of their country's regulatory team argued that changing the existing safety culture requires continuous work and cannot be achieved solely through paperwork:

> One of the offshoots, of course, of regulating or overregulating is the paperwork that's demanded each and every show, and pilots and paperwork do not mix well. The danger about insisting on minutiae in the paperwork is that the guys will gloss over it.

Then another air show performer added, "What we are doing now is just filling out more papers and more papers, and they think it would make it better. It does not work with paper. It has to be here [brain] and here [heart]."

A respondent suggested adopting a "flag behavior" approach, which involves developing an internal mental warning system that raises a "flag" when something seems wrong or deviates from the standard:

> We have to change. What we need, I call "flag behavior". It is like the old times with a flag falling into an instrument, but now this "flag" has to be in the brain. This flag has to come and say; you do something I do not like – something is wrong with it. This is a culture I would like to go to – that would be my aim if I had changed something. Also in myself, I always tried to use this flag and advised myself, "Oops, something wrong, one notchback – first, look what is happening". Some people do not have it; they think that "It is an air show now where everybody wants to be lower". It is human, but we have to make it clear in our brains and change our behavior patterns.

This mental "flag" can help individuals reassess their actions and maintain safety. The interviewee emphasized that this cultural change is necessary to improve behavior patterns and ensure a safer environment for air shows.

3.5.1.2 Ownership

In discussions about the air show community's culture, a process ownership issue emerged, emphasizing the critical need for responsibility and accountability among performers, organizers, and senior leadership (see Figure 3.3).

3.5.1.2.1 Air Show Performers

Air show performers stated that as part of the need for responsibility and accountability in the air show community, personality traits such as discipline and humility should be desired, with humility being essential for making safety-critical decisions even in high-risk situations. A respondent mentioned that pilots from renowned teams like the Blue Angels, Thunderbirds, and Snowbirds were humble and kind, reflecting the importance of these personality attributes.

Respondents also noted that air show performers should exhibit professionalism, which includes being open to constructive criticism. As one civilian performer noted, competition pilots are judged on numerous aspects, so accepting critique and flying correctly are essential for growth.

Resilience was also identified as a desirable trait for air show performers. Some respondents with significant military demonstration team experience emphasized that air show performers must learn to anticipate and respond to unexpected situations. They should always be aware that inadvertent errors and mistakes can affect the performance of exceptional pilots. Another respondent agreed that adaptability in handling unexpected situations is a crucial skill for air show performers.

3.5.1.2.2 Existing Leadership in the Air Show Community

The air show performers discussed the existing leadership in the air show community, emphasizing the roles of air bosses, air show councils, and aviation authorities, as well as personal responsibility. An effective air boss should possess traits such as confidence, good planning, control, and a high-level situational awareness to ensure a smooth and safe air show.

Air show councils, particularly the European Airshow Council (EAC) and the International Council of Air Shows (ICAS), play vital roles in education and promoting safe attitudes for pilots, display directors, and regulators. The ICAS, in particular, has been acknowledged for creating the ACE system for granting display licenses and its philanthropic role. One of the most critical functions of air show councils is organizing annual conventions and providing networking and experience-sharing opportunities for air show performers.

3.5.1.2.3 Pilot's Selection Criteria

Respondents emphasized the crucial role and functions of various aviation regulatory authorities, top leadership, and military command top hierarchy in selecting air show pilots. One air show performer with military demonstration team experience mentioned that they mitigate risk by using strict selection criteria, ensuring that candidates have a fighter or ejection seat background, and often have flight instructional experience. This approach provides air show pilots with the most amount of high performance, high speed, and ejection seat time possible. A civilian respondent also stressed that air show safety culture and, subsequently, safer operations begin with the recruitment of pilots, followed by high-quality training, conducive living and sleeping conditions, and finally, the setup of a display.

3.5.1.2.4 Personal Responsibility

Most air show performers agreed on the importance of personal responsibility and leadership skills for air show performers. One respondent mentioned that while the air show fraternity can do much, it ultimately comes down to individual pilots' actions in a high-risk environment. Another emphasized the significance of performing safely to avoid reputational damage and its potential adverse financial impact.

The majority of respondents confirmed that all air show community stakeholders share responsibility and accountability for establishing an effective safety culture. One North American air show performer described the structured nature of air shows, emphasizing the focus on security and safety for all involved. The performer also noted that regardless of an air show's size, standardized safety requirements are followed in their country.

3.5.1.3 Continuous Enhancements

The need for continuous monitoring of any safety system for operational drift is essential for continuous enhancements. Such periodic improvements are also key in sustaining a proactive, robust, and resilient safety culture (Hollnagel, 2010; ICAO, 2018). Respondents discussed and affirmed a need for continuous enhancement of organizational processes, policies, procedures, and individual practices to maintain a resilient safety culture in the air show industry (see Figure 3.3).

3.5.1.3.1 Individual-Level Commitments

Several respondents agreed that at the individual level, air show performers should consider adding altitude margins to improve their flying, while others stressed the importance of creating margins for error to react to distractions or unforeseen events during demonstrations in high-threat, low-level aerobatic environments.

One interviewee stated:

> No one talks about [altitude] margins now. We forced the conversation now, sort of like a buzzword that everybody is accepting, but no one talked about margins until about five years ago. And so, you got minimums, but that is your minimum not to die. What is your margin for an air show so that you do not even think about it, and a lot of people do not have those.

Another interviewee added, commenting:

> Just because you are clear to 100 feet or 250 feet for rolling maneuvers – it does not mean to say you have to fly at that height; it is the basic rule below which you must

> not come. So why fly at it, because any minor distraction could put you underneath that height. So, it is always best to leave a little bit of a cushion. But you do get these pilots who determinedly fly down to the limit because they can, and they are the guys I worry about.

Some respondents also emphasized that individual air show performers should monitor and control inflated egos or an extreme sense of self-esteem. From a behavioral perspective, such an extreme sense of self-esteem or self-importance can lead to poor bonding with colleagues and team dynamics. Such inflated egos are potential harbingers for operating outside the established rules (violations). A characteristic statement by one interviewee was, "You cannot have so many pilots – with individual goals and individual egos – egos are checked at the door."

3.5.1.3.2 Organizational-Level Commitments

Several respondents discussed the importance of continuous improvement at the organizational level and mentioned the four pillars of the safety management system (SMS), namely safety policy and objectives, safety risk management, safety assurances, and safety promotions, as an effective performance-based tool to enhance safety in the international air show community. Despite the touted benefits of SMS implementation, some respondents expressed concern about overprescriptive safety regulations and the associated paperwork burden in their nation's air show sectors, arguing that such excessive documentation of processes can lead to "pencil whipping" and complacency and may not effectively improve safety.

Respondents also suggested that the air show community could learn from other high-reliability organizations (HROs) like nuclear power plants, firefighting units, and air traffic control experiences fewer than anticipated accidents or events of harm despite operating in highly complex, high-risk environments where even small errors can lead to tragic results. In the aviation sector, the Red Bull AIr Races were highlighted. A respondent noted the impressive safety record of these races, which achieved almost 100 races without an accident, noting that despite some close calls, there had been no fatalities.

Another respondent cited Formula 1 racing as another sector where lessons can be learned. The respondent noted that in Formula 1 racing, impressive safety records had been achieved through the use of new safety technologies, such as the halo safety device on cockpits, as well as protective equipment like fire-resistant coveralls and crash-resistant helmets. The air show community could potentially benefit from emulating these safety advancements to enhance performer safety.

3.5.2 Effect of Air Show Flying Experience on Perception of Resilient Safety Culture

Regardless of operational experience in the air show community, the maintenance of a resilient safety culture is key to continuous safety improvements in the air show industry. The perspectives of air show performers were sought in the study on how flying experiences impact resilient safety culture in the air show sector. Most respondents intimated that the higher the experience and exposure of an air show performer, the higher their expectations in terms of safety from the air show community. Experienced air show performers expect high standards from the rest of the display pilots and the management of air shows, specifically from the air bosses.

Respondents noted that most highly experienced air show performers have built-in resilience skills that help them anticipate the unexpected and manage any distractions and interruptions before and during the display. Paradoxically, some of these highly experienced pilots, especially those who fly solo twice or three times per day in air shows, might end up being complacent and conduct hazardous display profiles that could harm themselves and the entire air show community.

On the contrary, some respondents noted that inexperienced air show performers could sometimes overestimate their skills and capabilities, which is an intrinsic threat with safety risk (Dunning, 2011). To ensure a culture of excellence and resilience, as discussed earlier, these novice air show performers must be mentored by more experienced and safety-conscious air show performers. Experienced, respected, and safety-conscious mentors should be willing to share their experiences and learning outcomes accumulated over the years in the air show display business with others to engender the transfer and retention of organizational knowledge.

Nevertheless, irrespective of the level of air show flying experience, the findings of this study suggested that most air show performers are inclined to accept and embrace a resilient safety culture that emphasizes best practices across the international air show community. It will be valuable for the more experienced air show performers with leadership or supervisory roles, such as instructors, mentors, subject matter experts, and evaluators, to spearhead the culture of proactive and resilient safety in the air show community.

3.6 CASE STUDY: FAST JET SOLO DISPLAY PILOT SAFETY WORKSHOP

3.6.1 BACKGROUND

The Fast Jet Solo Display Pilot Safety Workshop, conceived in 2014, is a one-of-a-kind international gathering for individuals responsible for showcasing their nation's military frontline fast jet aircraft.

Initially created exclusively for the F-16 "Viper" community, due to its prevalence in the European Theatre at that time, the workshop's value was soon recognized and expanded to include all nations with fast jet displays and further broadened in 2019 to encompass turboprop aircraft displays as well.

Open to current fast jet military solo display pilots, the meeting aims to facilitate the exchange of display experiences, develop safety standards for both displays and display training, and encourage discussions on key issues affecting participants in their roles. Issues related to *human factors* and *technology* in ensuring operational safety are consistently emphasized each year.

The workshop also provides an invaluable opportunity for incoming display pilots to connect with their peers before the upcoming season and develop both professional and personal relationships. By participating in the convention, pilots and organizers gain wider exposure to the European air show community, enabling them to meet in person before the season and plan well in advance.

3.6.2 Workshop's Exclusive Features Shaping Display Pilots' Safety Culture

The Fast Jet Solo Display Pilot Safety Workshop features several exclusive characteristics that contribute to shaping the safety culture of the participant display pilots. The following sections outline how each of these characteristics affects the air show safety culture:

- Invitation Only

By making the workshop invitation only, the organizers ensure that the attending pilots are highly skilled and experienced in fast jet solo aerobatic displays. This selection process results in a focused group of participants who share common goals and a commitment to safety. It creates an environment where pilots can openly discuss challenges and share best practices, knowing that their peers understand the unique nature of their profession. This exclusivity also fosters a sense of responsibility and pride among the attending pilots, as they recognize their role in promoting and maintaining a safety culture within their community.

- Same Training Background: Military Pilots Only

Limiting the workshop to military pilots ensures that all participants share a common working ethos, similar training background, and a high level of professionalism. Military pilots typically undergo rigorous training programs that emphasize safety, discipline, and adherence to established procedures. By bringing together pilots with this shared background, the workshop encourages the exchange of ideas, experiences, and best practices that are directly applicable to their unique context. It also allows participants to develop a deeper understanding of safety culture from the military perspective, which can be applied to improve safety in fast jet solo aerobatic displays.

- Connection Throughout the Year

Maintaining a connection among workshop participants throughout the year using messaging platforms fosters a sense of community and ongoing collaboration. This continuous communication allows pilots to share updates, discuss safety-related issues, and seek advice from their peers in real time. As a result, pilots can stay informed about emerging safety concerns, new best practices, and lessons learned from incidents or near misses. This ongoing dialogue contributes to the reinforcement of safety culture among the participating pilots and ensures that safety remains a top priority throughout the year.

- Closed-Door Session

Holding closed-door sessions at the workshop promotes an environment of trust and confidentiality, enabling pilots to openly discuss sensitive topics and share their experiences without fear of negative repercussions. By creating a safe space for honest conversations, the workshop encourages pilots to address safety concerns, admit mistakes, and learn from their peers. This open dialogue contributes to the development of a just culture, which is essential for fostering a strong safety culture. Pilots are more likely to report incidents and near misses, knowing that their peers will focus on learning and improvement rather than assigning blame.

3.6.3 Recommendations for the Air Show Community

Based on the lessons and best practices from the Fast Jet Solo Display Pilot Safety Workshop, the following recommendations can be made for the broader air show community to enhance the existing air show culture:

- Conduct Specialized Safety Workshops

Organize similar workshops tailored to specific types of display pilots, air show organizers, regulators, and air bosses, focusing on unique challenges and risks associated with their roles. This will ensure that all stakeholders have access to specialized training that enhances their understanding of safety culture and risk management.

- Establish a Culture of Open Communication

Encourage open dialogue and information sharing among air show stakeholders, allowing them to learn from each other's experiences and best practices. This can be facilitated through forums, workshops, or online platforms dedicated to discussing safety-related issues.

- Create Mentorship Programs

Develop mentorship initiatives that pair experienced professionals with those who are new to the industry, fostering knowledge sharing and guidance that can enhance safety culture and performance.

- Promote Safety as a Shared Responsibility

Emphasize the importance of collaboration and cooperation among all stakeholders, recognizing that safety is a shared responsibility. Encourage joint problem-solving to address safety concerns and improve the overall safety culture.

The Fast Jet Solo Display Pilot Safety Workshop offers valuable insights and best practices to the participating pilots that can be adopted by the broader air show community to enhance safety culture and excellence. By implementing these recommendations, display pilots, air show organizers, regulators, and air bosses can contribute to a safer and more successful air show environment for all involved. The ongoing commitment of all stakeholders to prioritize safety and excellence is essential for the continued growth and prosperity of the air show industry.

This concept of sharing knowledge and fostering collaboration can be extended to other high-performance sports and excellence-related industries. High-reliability organizations (HROs) like nuclear power plants, firefighting units, and air traffic control could benefit from such events by sharing best practices and refining operational standards. High-performance sports like Formula 1 racing, professional cycling, and competitive sailing could also adopt similar gatherings to improve safety and performance.

NOTE

1. A Ribbon Cut is an aerial maneuver performed by airshow pilots, which involves cutting a ribbon or a streamer suspended between two poles using the aircraft's wings, propeller, or tail. The ribbon is typically placed at a low altitude, often between 15 tand 25 feet above the ground. The objective of the maneuver is to demonstrate the pilot's precision flying skills, aircraft control, and mastery of low-altitude flying.

   During a Ribbon Cut performance, the pilot must approach the ribbon with great accuracy, often flying inverted or at a specific angle, to cut the ribbon without damaging the aircraft or compromising safety. This maneuver is visually captivating for the audience, showcasing the pilot's skill and daring, and is a popular staple at many airshows.

LIST OF REFERENCES

Adjekum, D. K., & Fernandez Tous, M. (2020). Assessing the relationship between organizational management factors and a resilient safety culture in a collegiate aviation program with Safety Management Systems (SMS). *Safety Science*, *131*, 104909. https://doi.org/10.1016/j.ssci.2020.104909

Akselsson, R., Koornneef, F., Stewart, S., & Ward, M. (2009a). Resilience Safety culture in aviation organisations. In *HILAS: Human integration in the lifecycle of aviation systems*, 20.

Akselsson, R., Koornneef, F., Stewart, S., & Ward, M. (2009b). *Resilience safety culture in aviation organisations*. https://www.researchgate.net/publication/254908155_Resilience_Safety_Culture_in_Aviation_Organisations

Cantu, J., Tolk, J., Fritts, S., & Gharehyakheh, A. (2020). High Reliability Organization (HRO) systematic literature review: Discovery of culture as a foundational hallmark. *Journal of Contingencies and Crisis Management*, *28*(4), 399–410. https://doi.org/10.1111/1468-5973.12293

Clarke, S. (2000). Safety culture: Under-specified and overrated? *International Journal of Management Reviews*, *2*(1), 65–90. https://doi.org/10.1111/1468-2370.00031

Dunning, D. (2011). The Dunning–Kruger effect: On being ignorant of one's own ignorance. In J. M. Olson & M. P. Zanna (Eds.), *Advances in experimental social psychology* (Vol. 44, pp. 247–296). Academic Press. https://doi.org/10.1016/B978-0-12-385522-0.00005-6

Federal Aviation Administration. (2022). *Risk management handbook (FAA-H-8083-2A)*. Federal Aviation Administration (FAA). https://www.faa.gov/regulationspolicies/handbooksmanuals/risk-management-handbook-faa-h-8083-2a

Glendon, I., & Stanton, N. (2000). Perspectives on safety culture. *Safety Science*, *34*(1–3), 193–214. https://doi.org/10.1016/S0925-7535(00)00013-8

Harris, D. (2011). *Human performance in the flight deck*. CRC Press.

Heese, M., Kallus, W., & Kolodej, C. (2014). Assessing behaviour towards organizational resilience in aviation. *Proceedings of the 5th resilience engineering association symposium* (pp. 67–73).

Hofstede, G., Hofstede, G. J., & Minkov, M. (2010). *Cultures and organizations: Software of the mind; intercultural cooperation and its importance for survival* (3rd ed.). McGraw-Hill.

Hollnagel, E. (2010). How resilient is your organisation? An introduction to the Resilience Analysis Grid (RAG). *Sustainable transformation: Building a resilient organization*. https://hal-mines-paristech.archives-ouvertes.fr/hal-00613986

Hollnagel, E. (2014). *Safety-I and safety-II: The past and future of safety management*. CRC Press.

Hollnagel, E., Pariès, J., Woods, D. D., & Wreathall, J. (2011). *Resilience engineering in practice: A guidebook*. Ashgate Publishing, Ltd.

ICAO. (2018). *Doc 9859, safety management manual*. International Civil Aviation Organization (ICAO). https://skybrary.aero/sites/default/files/bookshelf/5863.pdf

Liao, M.-Y. (2015). Safety culture in commercial aviation: Differences in perspective between Chinese and Western pilots. *Safety Science*, *79*, 193–205. https://doi.org/10.1016/j.ssci.2015.05.011

Mearns, K. J., & Flin, R. (2018). Assessing the state of organizational safety-culture or climate? In *Validation in psychology* (pp. 5–20). https://doi.org/10.4324/9781351300247-1

Merritt, A., & Helmreich, R. L. (1996). Creating and sustaining a safety culture: Some practical strategies (in aviation). In *Applied aviation psychology – Achievement, change, and challenge* (pp. 20–26).

Patankar, M. S., Brown, J. P., Sabin, E. J., & Bigda-Peyton, T. G. (2012). *Safety culture: Building and sustaining a cultural change in aviation and healthcare*. Ashgate Publishing, Ltd.

Pillay, M., Borys, D., Else, D., & Tuck, M. (2010). *Safety culture and resilience engineering: Theory and application in improving gold mining safety* (pp. 129–140).

Reason, J. (1997). *Managing the risks of organizational accidents* (2nd ed.). Routledge.

Reason, J. (2016). *The human contribution*. Routledge.

Schein, E. H., & Schein, P. A. (2016). *Organizational culture and leadership* (5th ed.). Wiley. https://www.wiley.com/en-us/Organizational+Culture+and+Leadership%2C+5th+Edition-p-9781119212041

Shirali, Gh. A., Motamedzade, M., Mohammadfam, I., Ebrahimipour, V., & Moghimbeigi, A. (2015). Assessment of resilience engineering factors based on system properties in a process industry. *Cognition, Technology & Work*, *18*. https://doi.org/10.1007/s10111-015-0343-1

Shirali, Gh. A., Shekari, M., & Angali, K. A. (2016). Quantitative assessment of resilience safety culture using principal components analysis and numerical taxonomy: A case study in a petrochemical plant. *Journal of Loss Prevention in the Process Industries*, *40*, 277–284. https://doi.org/10.1016/j.jlp.2016.01.007

Smith, A. F., & Arfanis, K. (2013). 'Sixth sense' for patient safety. *British Journal of Anaesthesia, 110*(2), 167–169. https://doi.org/10.1093/bja/aes473

Sujan, M. A., Furniss, D., Anderson, J., Braithwaite, J., & Hollnagel, E. (2019). Resilient health care as the basis for teaching patient safety – A safety-II critique of the World Health Organisation patient safety curriculum. *Safety Science, 118*, 15–21. https://doi.org/10.1016/j.ssci.2019.04.046

Teske, B. E., & Adjekum, D. K. (2022). Understanding the relationship between High reliability theory (HRT) of mindful organizing and safety management systems (SMS) within the aerospace industry: A cross-sectional quantitative assessment. *Journal of Safety Science and Resilience, 3*(2), 105–114. https://doi.org/10.1016/j.jnlssr.2022.01.002

UK Civil Aviation Authority. (2023). *CAP 403: Flying displays and special events: Safety and administrative requirements and guidance.* UK Civil Aviation Authority. https://publicapps.caa.co.uk/docs/33/CAP403%20Edition%2020.pdf

United Kingdom Health and Safety Executive. (2002). *Evaluating the effectiveness of the health and safety executive's health and safety climate survey tool.* (Research Report No. 042; p. 54). Health and Safety Executive.

United States Nuclear Regulatory Commission. (2020, June 8). *Safety culture.* United States Nuclear Regulatory Commission. https://www.nrc.gov/about-nrc/safety-culture.html

Weick, K. E., & Sutcliffe, K. M. (2001). *Managing the unexpected: Assuring high performance in an age of complexity* (pp. xvi, 200). Jossey-Bass.

Weick, K. E., & Sutcliffe, K. M. (2009). Managing the unexpected: Resilient performance in an age of uncertainty. *Personnel Psychology, 62*(3), 646–652. https://doi.org/10.1111/j.1744-6570.2009.01152_6.x

Yorio, P. L., Edwards, J., & Hoeneveld, D. (2019). Safety culture across cultures. *Safety Science, 120*, 402–410. https://doi.org/10.1016/j.ssci.2019.07.021

4 Psychological Factors

4.1 INTRODUCTION

Anecdotally, there seems to be a high public perception that air show performers are adrenaline "junkies" who engage in high-risk and thrilling aerial displays, pushing themselves to the limits of what is humanly possible. While these performers may appear fearless and daring, the psychology behind their performances is much more complex. A search of extant literature suggests a paucity of discourses and understanding of several psychological factors that shape the mindsets of air show performers within the broader air show community. Despite the noted scarcity of research, it is important to understand the psychological challenges faced by these performers in their pursuit of excellence and explore strategies needed to sustain stellar performance within the limitations of such challenges. By examining air show performers' mental preparation, attentional resources, and resilience, we provide a unique perspective on the often-hidden psychological factors behind these spectacular performances during air shows.

The psychological challenges faced by air show performers are immense during displays. Pilots must cope with high-performance standards and expectations, leading to extreme stress and anxiety. Pilots have to deal with attentional issues that can impact the quality of decisions, most of which are intuitive and time-critical. These decisions can also have life-or-death consequences while operating in a hostile low-level environment.

We do not delve into the generic complexity of conceptual frameworks in aviation psychology in this Chapter. We focus on critical psychological factors that practically impact air show performers, such as hazardous attitudes and inherent biases. We also explore the impact of gravitational forces (G Forces) and low-level aerobatics on the psychological capabilities of air show performers. We then proffer plausible mitigations to these hazardous attitudes and biases. We conclude this chapter with an analysis of the relevant psychological factors that contributed to the Shoreham tragedy in 2015.

The psychological aspects of air show performances may have been underestimated by the air show community, leading to a lack of awareness of how these factors affect the performance of pilots and other stakeholders in the air show industry. This chapter provides some beneficial information for all stakeholders on psychological factors that impact air show performance and also for aviation accident investigators during the review of air show mishaps. Based on the Human Factors Analysis and Classification Systems (HFACS) taxonomy by Wiegmann and Shapell (2000), some investigators may regard "pilot error" as an insufficient explanation for an accident and may seek to understand the psychological factors underlying a pilot's

DOI: 10.1201/9781003431879-4

decision-making process. That is why a discussion on psychological factors impacting air show performance is apropos in this book.

By discussing issues associated with these factors, we can also enhance the air show community's awareness of human psychological vulnerabilities and the potential impact on decision-making. This can invariably lead to the adoption of policies and processes that considers these vulnerabilities and assure safer air show performances industry-wide.

4.2 HAZARDOUS ATTITUDES

A psychological factor suggested to impact decision-making and be a catalyst for accident involvement is *hazardous attitudes* (Martinussen & Hunter, 2018). Regulatory guidance from the Federal Aviation Administration (FAA) suggests five critical attitudes that can lead to hazardous events: macho, antiauthority, invulnerability, impulsivity, and resignation (FAA, 2016). Therefore, hazardous attitudes in any air show organization need to be expeditiously identified, and the risk associated with it mitigated to a negligible effect using a risk matrix or guidelines such as the FAA antidotes to hazardous attitudes (FAA, 2016).

The FAA guidance provides a set of antidotes to mitigate the adverse effects of the five hazardous attitudes identified among pilots: A macho attitude can be mitigated with a "taking chances is foolish" approach; an antiauthority attitude needs a "follow the rules—they are usually right" method of correction; invulnerability requires an "it could happen to me" approach; while impulsivity needs the pilot to slow down and "not [act] so fast. Think first"; and, finally, pilots who have a resigned attitude should be supported with an approach that says, "I am not helpless. I can make a difference" (Federal Aviation Administration, 2016, pp. 2–5).

A hazardous attitude can lead to unsafe behavior, adversely affecting the display pilot's performance during air shows. Barker (2020, p. 622) defined a *rogue display pilot* as "an unprincipled pilot living apart from the display community and having destructive tendencies." According to Barker, rogue pilots are characterized by hazardous attitudes such as placing their ego above all they do: They push the boundaries and limits with aggression and arrogance in their ignorance – which could be attributed to the Dunning-Kruger effect[1] (Dunning, 2011; Kruger & Dunning, 1999) – and they risk not only their lives but also the lives of other pilots, and the air show spectators.

The B-52 accident at Fairchild Air Force Base in 1994 has been identified as an example of a rogue aviator incident (Barker, 2003, 2020; Diehl, 2002; Kern, 1995; Thompson, 1999; U.S. Air Force, 1994). Therefore, to prevent or minimize such egocentric and "rogue" attitudes, it is critical that the air bosses demonstrate effective leadership during the planning and execution phase of the air show. Roger Beazley, former flying display director of the Farnborough International Air show, suggested a leadership approach that air bosses could take to identify and mitigate any egocentric and "rogue" attitudes by the air show performers when he suggested that an aircrew safety briefing should be the forum to set the tone for safer and responsible behaviors. He intimated that "air show briefing works wonders in clearing the air and starting the new day fresh" by ensuring "an open and free debate, in private, on previous days' problems involving all participants" (Barker, 2020, p. 50).

4.3 BIASES

In psychology, a bias is a systematic error in judgment or decision-making arising from cognitive processes rather than random chance or external influences (Fiske & Taylor, 2021). Biases often result from mental shortcuts or heuristics that individuals use to simplify complex information or to make decisions more quickly (Tversky & Kahneman, 1974). While these shortcuts can be efficient in certain situations, they can also lead to systematic errors and flawed judgments.

Many types of biases have been identified in psychological research, such as plan continuation, confirmation, overconfidence or illusion of superiority, confirmation, anchoring, and social facilitation. These biases can affect various aspects of human cognition, including perception, memory, and decision-making.

4.3.1 The 5B Model

Numerous cognitive biases can influence aviators' decision-making, and air show performers are not exempt. Understanding and addressing these biases is essential to enhancing safety and performance in the air show industry.

Therefore, we introduce a *5B* approach that aims to assist air show performers in the identification and management of these biases and their effect on safe decision-making, both on the ground and in the air. This is particularly important for solo performers, such as solo display pilots, parachutists, wing walkers, and others who might not have peer support during their performances before large audiences.

4.3.1.1 Introduction

The most relevant biases for air show performers, as identified by the authors, include (see Figure 4.1):

4.3.1.1.1 Bias 1. Plan Continuation

This bias occurs when individuals persist in their original plan or course of action despite new information suggesting that an alternative might be more appropriate. Air show performers may be inclined to continue their performance, even in unfavorable conditions, leading to potentially dangerous situations. Dekker (2017) suggests that this bias normally occurs when there is a natural tendency that favors not changing plans or individuals are unwilling to change plans under high workload, high-time pressure, or unspecified problem situation. It can also be prevalent when individuals perceive it as a waste of cognitive investments and physical energy up to the point of requiring a change of plan.

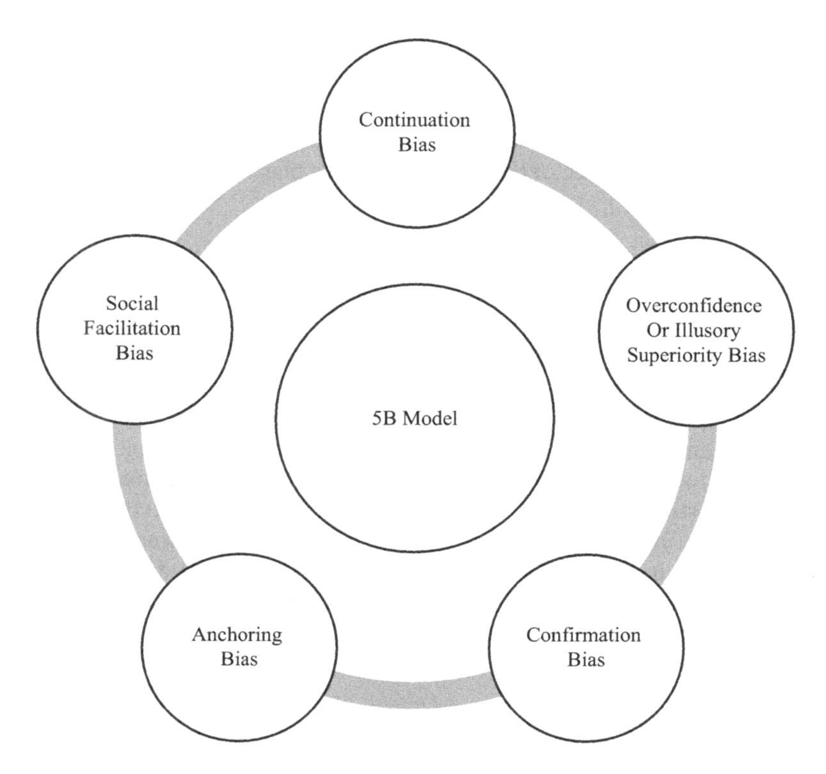

FIGURE 4.1 The *5B* model.

In an imaginary scenario, "Mike," a skilled air show performer with his Pitts Special, a popular aerobatic biplane, experiences plan continuation bias during his performance at a popular air show. Despite worsening weather conditions, including thunderstorms and strong winds, Mike insists on completing his planned routine. This determination to stick to the original plan endangers his safety and, potentially, that of the spectators. Eventually, after losing control momentarily, he realizes the risk and aborts the performance. The scenario highlights the importance of being flexible and adaptive in decision-making, prioritizing safety over adhering to a predetermined plan.

4.3.1.1.2 Bias 2. Overconfidence and Illusory Superiority

Overconfidence refers to an individual's tendency to overestimate their abilities or knowledge, while illusory superiority is the belief that one's skills are above average (Dunning, 2011; Kruger & Dunning, 1999). This can lead to an inflated perception of the accuracy of one's knowledge and justified actions. Air show performers may fall prey to these biases, leading to excessive risk-taking or inadequate preparation.

In a hypothetical scenario, "Bruce" is an overconfident fast jet display pilot performing at a famous air show. Bruce's overconfidence and illusory superiority lead him to attempt a risky maneuver called the "Lightning Spiral," which he has only practiced a few times. During the maneuver, he misjudges the altitude and nearly crashes but manages to regain control just in time. The scenario highlights the potential dangers of overconfidence and illusory superiority in air show performances, emphasizing the importance of being aware of one's biases and limitations to ensure safety and success.

4.3.1.1.3 Bias 3. Confirmation

This bias occurs when people search for or interpret information in a way that confirms their pre-existing beliefs or expectations. Performers may selectively focus on evidence that supports their belief in their ability to execute a maneuver successfully while ignoring warning signs or potential hazards.

In a hypothetical scenario, a renowned air show performer, "Joe," is flying an Extra 300 at an international air show. Joe's confidence in his ability to execute his aerial routine is influenced by confirmation bias, as he selectively focuses on evidence supporting his belief in his skills while ignoring potential hazards. During his performance, Joe disregards changing wind conditions, leading to a dangerous situation in which he nearly crashes while attempting a low-altitude knife-edge pass. He narrowly avoids disaster and completes the performance. This scenario illustrates the potential consequences of confirmation bias and underscores the importance of considering all relevant factors and being aware of one's biases when performing high-risk activities.

4.3.1.1.4 Bias 4. Anchoring

Anchoring refers to the tendency to rely heavily on an initial piece of information (the "anchor") when making decisions. Air show performers may become anchored to a specific plan or strategy, which could limit their ability to adapt to changing conditions or new information.

In the hypothetical scenario, "Becky," a talented air show pilot flying the MXS-R, a modern and agile aerobatic monoplane, falls prey to anchoring bias during her performance at a local air show. Despite changing weather conditions and unpredictable wind gusts, she remains anchored to her original plan and strategy, unwilling to adjust her maneuvers. This leads to a dangerous situation when attempting a low-altitude inverted pass, but she narrowly avoids disaster. The scenario underlines the importance of being flexible and responsive to new information rather than becoming anchored to an initial plan or strategy in high-risk activities like air show performances.

4.3.1.1.5 Bias 5. Social Facilitation

Air show display pilots may sometimes engage in unsafe behaviors without any previous indicators of hazardous attitudes, primarily due to the underlying impact of social facilitation bias (Papadakis, 2008). Social facilitation bias refers to the phenomenon where an individual's performance is influenced by the presence of others, which can lead to alterations in their usual behavior. This bias can explain the tendency for individuals to perform better on simple tasks and worse on complex tasks when in the presence of others. The effect of social facilitation bias might be so powerful that it prevents a display pilot from recognizing a hazardous situation as it develops, ultimately putting their safety and that of others at risk.

In the hypothetical scenario, "Jean," a demo team leader and experienced air show pilot, is influenced by social facilitation bias during his performance at an international air show. The pressure of performing in front of an audience leads Jean to take more risks than usual, attempting a complex, risky maneuver called the "Falcon's Dive" in tight formation. The presence of the audience and the pressure to perform well affect the team's precision, leading to an unstable formation. Fortunately, the team members recognized the danger in time and aborted the maneuver, avoiding potential disaster. The scenario highlights the importance of display pilots remaining aware of the potential impact of social facilitation bias on their decision-making and performance, ensuring the safety of themselves and others.

It can be noted that air show performers performing in front of large audiences may experience heightened stress and anxiety to please the crowd, which could impair their decision-making and performance. Furthermore, Barker (2020) suggests that the significant discrepancy between accidents during actual air shows and practice sessions can be attributed to various factors, one of which is the pressure induced by the existence of spectators. The presence of an audience may cause pilots to push themselves beyond their usual limits in an attempt to impress, leading to an increased likelihood of accidents. In addition to the presence of spectators, other factors contributing to this disparity could include the unique conditions and environment of an actual air show, such as varying weather patterns, the need to adhere to tight schedules, and the challenge of coordinating multiple performances.

4.3.2 The *5B* Domino Effect

The domino effect of five specific cognitive biases – social facilitation, continuation bias, overconfidence, anchoring, and confirmation bias – could dramatically impact an air show performer's cognitive functions, leading to potentially hazardous consequences (see Figure 4.2). Social facilitation, the tendency to perform well-practiced tasks better when in the presence of others, could encourage the performer to push beyond their limits to impress the crowd, starting the chain of dominos.

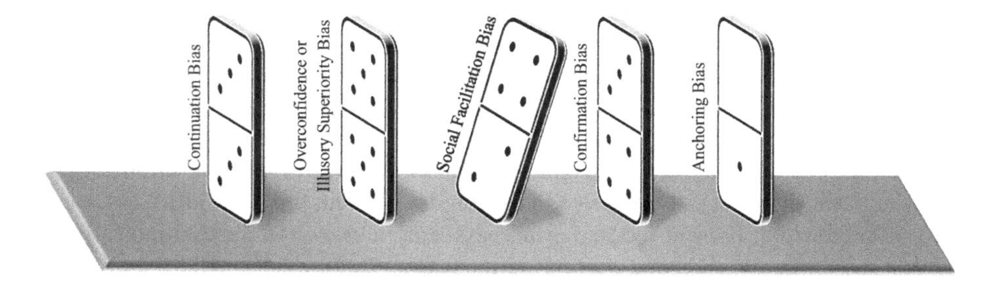

FIGURE 4.2 The *5B* domino effect.

This might then trigger the continuation bias, the irrational desire to continue an already initiated task, making the performer persist in a potentially dangerous stunt despite emerging risks. Overconfidence bias could exacerbate this situation, as the performer, driven by an inflated belief in their abilities, might underestimate these risks. The anchoring bias, a reliance on initial pieces of information to make decisions, could come into play if the performer is overly influenced by a successful early maneuver, leading them to think all subsequent stunts will go equally well. Finally, confirmation bias, the tendency to focus on information that confirms pre-existing beliefs, could prevent the performer from adequately responding to new information, indicating increased danger. Each bias acts as a domino, setting off the next, and together they could lead to a critical failure in the performer's cognitive function, increasing the likelihood of an accident during the air show.

Nevertheless, even a single cognitive bias could be enough to significantly impair an air show performer's decision-making capabilities, potentially leading to dangerous outcomes. For instance, overconfidence bias, where an individual has an exaggerated faith in their own skills or control over situations, could be particularly harmful. If the performer succumbs to overconfidence, they may underestimate the risks associated with their stunts, assuming they can handle any situation that arises based on their skills alone. This could prevent them from taking necessary precautions or aborting a stunt when conditions change or when something doesn't go as planned. Consequently, even without the compounding influence of other biases, this single cognitive bias can distort the performer's risk assessment and decision-making processes, thereby increasing the likelihood of an accident during the air show. This highlights the powerful impact that even one cognitive bias can have on a person's cognitive function in high-stakes situations.

Mindfulness, with its focus on present-moment awareness and nonjudgmental acceptance, can serve as a powerful tool to disrupt the domino effect of cognitive biases on an individual's decision-making process. By cultivating mindfulness, an air show performer, for instance, could better recognize the onset of cognitive biases such as overconfidence or confirmation bias. Rather than getting swept up in the bias, mindfulness could help the performer acknowledge the bias without judgment and then let it pass without acting on it. In this way, mindfulness can help to disrupt the domino effect by preventing one bias from automatically triggering another. Furthermore, by promoting a heightened state of awareness, mindfulness can help the performer stay attuned to the present moment and the realities of their situation, thus enabling more accurate risk assessment and safer decision-making. Consequently, mindfulness can act as a cognitive "reset" button, helping to prevent the chain reaction of biases that could potentially lead to dangerous outcomes.

Just as a gust of wind can sweep across a chain of dominos, pausing their cascade and helping them reset to their upright positions, so can the practice of mindfulness halt the tumbling effect of cognitive biases on decision-making. Mindfulness, with its gentle breeze of present-focused awareness, can intervene in the domino rally of biases, steadying the tiles and preventing them from toppling one after the other. It brings a moment of calm and clarity amidst the rush of falling dominos, allowing each tile – representing a distinct cognitive bias – to reset to its neutral position. This prevents the biases from triggering each other, interrupting the cascade before it can lead to a potentially dangerous decision. Like the wind restoring balance to the dominos, mindfulness brings equilibrium back to the cognitive landscape, enabling clearer, safer, and more grounded decision-making.

4.3.2.1 Mitigation Strategies

The *5B* approach provides some realistic mitigations for air show performers in managing the adverse effects of the biases discussed earlier:

4.3.2.1.1 Bias 1. Plan Continuation

Under the high workload and stressful environments of an air show, display pilots need to be conscious of changes in operational parameters and their implications for successful and safe performance. The use of what-if scenarios and contingency planning for scenarios that are plausible during air show displays can be helpful in identifying changes in the operational environments that require reviews. Identifying these changes and critically analyzing such changes based on the most acceptable safety risk criteria is essential in determining whether or not a revised course of action is needed with respect to the original plan. These changes must be timely and decisive since, with an increasing workload, particularly in solo display scenarios, there can be limited mental capacity for information processing.

4.3.2.1.2 Bias 2. Overconfidence

Air show performers should critically think about the end state of their actions and what happens if a performance goes awfully wrong due to their overconfidence. An understanding that they may not always have total control over situations due to constraints beyond their capabilities may tamper the sense of overconfidence. Also, a healthy dose of fear can be an impetus to let performers be reflective on the potential for making bad, impulsive, and reckless decisions. Acceptance and adherence to feedback, constructive critiques, and reviews from peers can also be helpful in minimizing overconfidence among air show performers. It will be insightful for the air show community to cultivate a culture where professionals are apt to share their mistakes, listen to the opinions of others, and learn from each other's experiences to foster humility and respect.

4.3.2.1.3 Bias 3. Confirmation

It will be intuitive for air show performers to be aware of their personal bias and how it influences their decision-making. They need to consider all the evidence available when assessing a problem rather than just the evidence confirming their views. The perspectives of peers, and especially those who hold contradictory views on a solution to a problem, can provide objective reflections. The flexibility to update and change one's mind when presented with new and sometimes conflicting evidence can be very helpful during air show activities.

4.3.2.1.4 Bias 4. Anchoring

Air show performers should elicit information from multiple sources, if possible, during air show planning and displays. Gathering knowledge can help in making effective safety risk-based decisions that have minimal anchoring facts. An iterative and introspective assessment of the decision-making process using standard operating procedures (SOPs) and guidance materials can help identify potential anchors. Seeking the opinions of experts and peers, such as air bosses and team coaches, when making operational decisions can help to minimize the effects of this bias.

4.3.2.1.5 Bias 5. Social Facilitation

Studies have suggested that the presence of others increases arousal when completing a complex task and increases the speed of performance for a simple task but decreases the speed for a complex task (Rafaeli & Noy, 2002). That suggests that for very complex tasks such as aerobatic displays and routines, performers should rely on their training, personal limitations, and prebriefed display routine to minimize the effects of this bias. Air show performers should practice routines until they become natural (or the dominant response), which builds professional confidence. This minimizes the need to stretch the safety and operational envelope in front of an air show audience while competing in a team or as a solo performer. Air show performers must understand that meeting established safety goals should be more important in terms of performance than merely pleasing an audience or trying to out-compete others.

The *5B* approach provides other stakeholders in the air show industry, such as air bosses, organizers, and regulators, with a tool to address some of the biases inherent in air show operations. By acknowledging the risk associated with these biases and effectively using safety risk controls such as training programs, safety briefings, and regulatory guidelines, the deleterious effects of these biases can be minimized to a tolerable level.

Furthermore, stakeholders who recognize that these biases may influence their decision-making processes can help create a more supportive air show environment cognizant of the psychological limitations of the human operator. This proactive approach will require effective communication, collaboration, and support among all parties involved in air shows, ultimately contributing to safer and more successful performances. Finally, an awareness of the limitations due to these biases can also bring into focus and prioritize the broader issue of mental health and wellness in the air show industry.

4.4 STUDY FINDINGS

As part of our research, we posed the following question to air show performers during the interview and focus group sessions "What are the most common hazardous attitudes observed among air show performers?"

Interestingly, there was a perception among respondents that, based on a historical trend, the number of air show performers with noticeable hazardous attitudes seems to be decreasing compared to previous years. Nevertheless, occasional incidents and accidents suggest that these attitudes still play a role as contributing factors. The respondents noted that some of the hazardous attitudes, such as *invulnerability, impulsivity, machismo, and antiauthority*, could be found among some performers, with invulnerability being the most prevalent. Interestingly, *resignation* was rarely mentioned as an attitude present among air show performers.

Some of the respondents also came up with what they termed "concealed" hazardous attitudes. Ego emerged as an attitudinal factor that must be managed to prevent it from becoming a psychological hazard. The respondents defined egocentric air show performers as those who only prioritize themselves and their parochial goals, show off to peers, family, and friends, and jeopardize the safety of the broader air show community.

Respondents also emphasized that some air show performers will not disclose their inadequate technical knowledge and "stick rudder" skills for some display profiles. These nondisclosures may be due to egocentric reasons or, in some cases, fear of losing clout among peers. Such attitudes pose dangers to both individual air show operators and the community as a whole. Some respondents also cited cases where performers lacked or had inadequate air show experiences, had deficient flying skills, and lacked professional discipline by not conducting proper pre-flight preparation. The respondents noted these as behavioral outcomes of "concealed" hazardous attitudes. It was also intriguing that respondents noted that some air show performers, out of financial constraints, can accept significant risks and engage in hazardous flight displays.

Passive distraction during air show performances was another concealed hazardous attitude mentioned. This refers to performers who become easily distracted by emotional and family issues, the audience, and social media, which may compromise their ability to perform safely and effectively.

Disorganization was identified as another concealed attitude among air show performers. They noted that some air show performers are not meticulous in their preparation and execution of displays. The respondents noted that such an excessively casual approach to displays could lead to careless flying, which poses safety risks within the air show community.

In terms of recommendations to manage these hazardous attitudes, respondents suggested mentorship by show instructors and evaluators, the use of evidence-based educational programs for air show performers, and informal hangar sessions led by peers. Details of respondents' perspectives on hazardous attitudes are discussed in the next section.

4.4.1 Hazardous Attitudes Identified by the FAA as Detrimental to Flight Safety

In terms of assessing the severity of hazardous attitudes identified by the FAA as detrimental to flight safety, respondents ranked *invulnerability* as the most dangerous, followed by *macho*, *impulsivity*, and *antiauthority* in that order, with *resignation* not being observed as prevalent among air show performers (see Figure 4.3). Some respondents shared stories of performers who exhibited these hazardous attitudes, which unfortunately led to fatal accidents. One interviewee recalled a display pilot with a mix of hazardous attitudes, as stated below:

> *This guy had this macho attitude, combined with invulnerability, combined with anti-authority, where he was told, "Do not do anything funny." Instead, he was doing maneuvers that it was hard to understand that were possible in a full-size aircraft. He was doing them in his model aircraft; he was also a national RC model champion. So, he really could fly well, but he was trying to use that in a full-size aircraft, deep alpha stalls, and stuff for what reason? Those spectators would not have appreciated what he was doing, and he messed it up, and he paid with his life. Horrible situation.*

Numerous respondents associated invulnerability with complacency, resulting in performers exceeding their limitations, and one interviewee recounted:

> *Sadly, in the last five years or so, particularly in my country, most of the accidents and incidents have been by very experienced pilots, and that is the bit that worries me now. It could be because of complacency—they have flown the same airplane, the same routine, 24 times this season, so they just get in the airplane, and off they go, and they either forget to check or they use the wrong height.*

According to the same interviewee, this type of attitude could result in air show performers exceeding their limitations: "*But you do get these pilots who determinedly fly down to the limit because he can, and they are the guys I worry about.*"

A *macho* attitude, marked by overconfidence and overestimating one's skills and capabilities, was also noted as problematic by some respondents. Regarding the macho attitude, an interviewee recounted an instance of an air show performer who exhibited this type of behavior but tragically died in another accident, as described below:

> *He had that machismo, and he was sort of always out to prove something. He pushed an inverted maneuver too hard, well below where I thought or, given the conditions of the day, as it was a very windy day. I was surprised that they allowed him to fly that day; it was that windy. And he pushed an inverted maneuver too low, and he recovered within feet of hitting the ground.*

The *antiauthority* attitude, characterized by disregarding rules and regulations, was highlighted as extremely dangerous. One interviewee mentioned that the air show industry's mindset had shifted from accepting a "my right to die" mentality to being more focused on safety and adhering to regulations.

4.4.2 Concealed Hazardous Attitudes

As stated in the introductory summary, respondents also identified *concealed* hazardous attitudes among air show performers in addition to the nominal ones

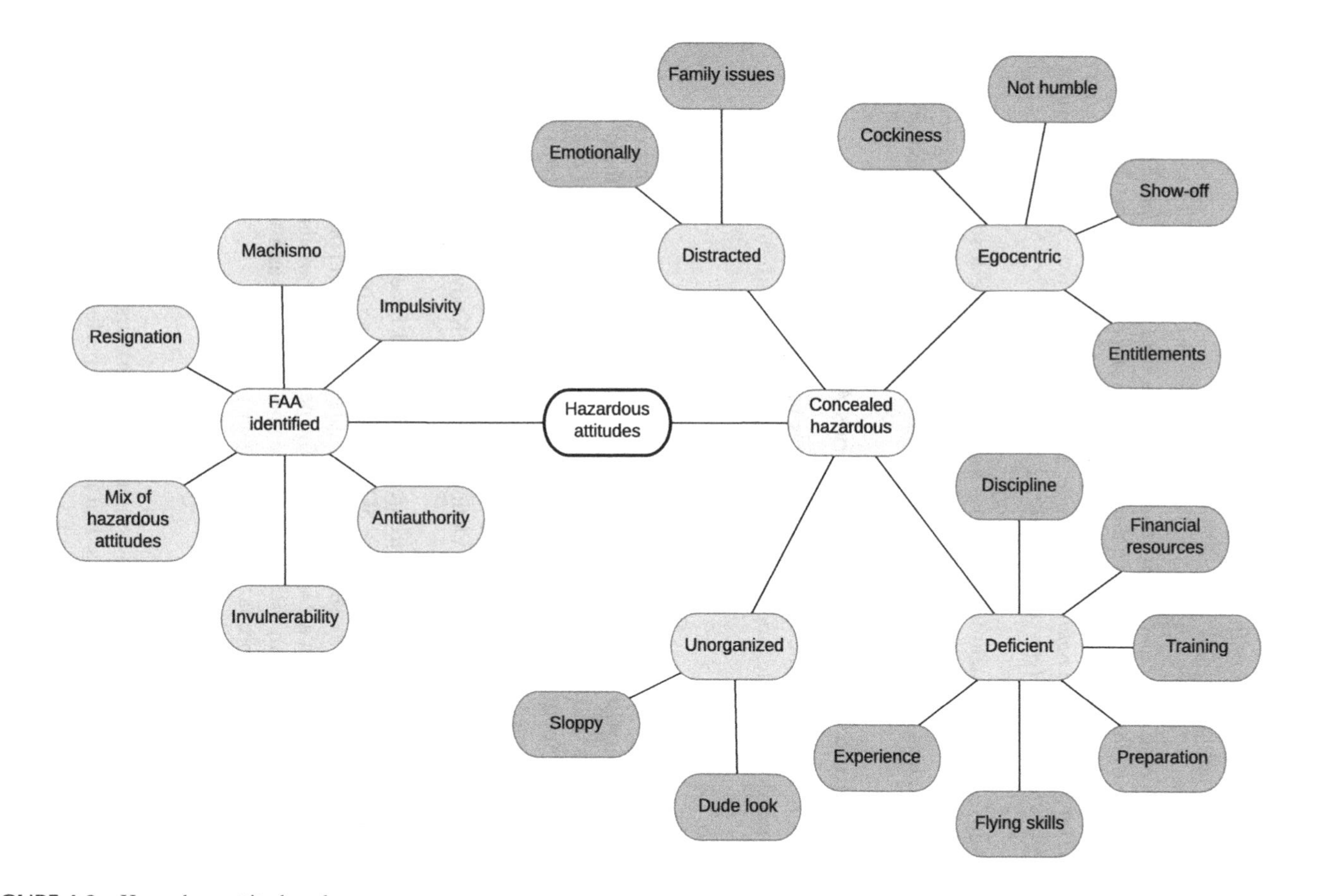

FIGURE 4.3 Hazardous attitudes, themes map.

identified by the FAA as detrimental to flight safety. These concealed hazardous attitudes include egocentrism, distraction, and disorganization (see Figure 4.3); an interviewee, who has vast experience performing at air shows, recounted an instance of an air show performer with such an attitude who tragically died in a fatal accident, as mentioned below:

> *He had this little surfer dude going, and I thought in the back of my mind I thought there is something it does not fit with air show pilots; it is not what you want to see around air shows and airplanes. It is cool, it is okay to go surfing, but you do not have that attitude when renting an airplane. Sure enough, next year, it takes a kid up, crashes, and dies, and I never forgot that.*

To sum up, concealed hazardous attitudes among air show performers can compromise the safety and success of air show events. Recognizing and providing antidotes for these attitudes is crucial in ensuring a safe and professional air show environment.

4.4.3 Antidotes to Concealed Hazardous Attitudes

Respondents also recommended some antidotes to minimize the effects of the concealed hazardous attitude, and these are outlined below as a useful guide:

- Ego

An antidote to ego is to prioritize collaboration and teamwork. For instance, performers can participate in safety meetings or training sessions to help them identify potential hazards and take necessary steps to mitigate them. They can also mentor and support one another and work together as part of the air show family to promote safety and excellence to the new generations of spectators.

- Lack of Interest in Improving Deficient Technical Knowledge and Skills

Strict training and experience requirements augmented by evidence-based assessment techniques stipulated in air show SOPs and regulatory requirements can ensure proficiency and currency. Those who fail to ensure currency and proficiency lose out on the opportunity to fly in air shows. Air show performers can also seek help from mentors or coaches who can provide guidance and support to assist them in improving their flying skills and preparing for air shows.

- Lack of Financial Resources

Prudent financial planning and investment by performers can minimize the pressures posed by financial constraints. If it is possible, performers undergoing known financial challenges such as bankruptcy and business collapse should be carefully monitored by peers for potential high safety risk acceptance. At-risk performers can also seek out sponsors or other funding sources to support their operations and ensure that they can operate safely.

- Antidote to Distraction

The antidote to distraction is to develop a focus on the present moment. Air show performers can use mindfulness techniques like meditation or visualization to stay focused and present during performances. They can also minimize distractions, such as limiting access to social media or other distractions during performance times, and adhere to the 60-minute rule[2] (International Council of Air Shows, 2012).

- Antidote to Being Unorganized

The antidote to this attitude is to prioritize organization and ensure meticulous preparation. Air show performers can create checklists and systems to ensure they are thoroughly prepared for performances. They can also work with mentors or coaches to develop strategies for staying organized and on task during performances. Disorganization primed by complacency can be managed by ensuring that no matter the familiarity with any air show display routine, the same systematic approach is used for planning and reviewing safety guidelines and SOP materials. Periodic use of regular safety checks and drills among teammates and peers can also be helpful in identifying routine operational slacks and ingraining a safety culture that emphasizes vigilance.

Using these antidotes, air show performers can promote a safety culture and mitigate hazardous attitudes that can lead to accidents and injuries.

4.5 PSYCHOLOGICAL EFFECTS OF PHYSIOLOGY ON AIR SHOW PERFORMERS

The intense physical demands of air show performances can challenge performers physically and have significant consequences on their psychological well-being. The G forces encountered during specific maneuvers may result in physical discomfort or even loss of consciousness, further affecting the performers' cognitive functions.

A crucial factor that influences the psychological well-being of air show performers and potentially modifies these factors on an upward or downward trend is the direct impact of various flying parameters. These parameters include positive and negative G forces, high roll rates, rapid rates of descent, the sensation of air during wing walking, and changing formation parameters throughout the flight. The physiological effects on the human body due to these factors can subsequently alter the psychological status of an air show performer (Wickens et al., 2015).

Some studies suggest that high G forces affect cognitive functions: one study suggests that high G forces impair cognitive functions such as sensing, perception, reaction time, and decision-making (Banks et al., 2011), while a study (Green, 2016) that focused in long duration acceleration discovered that under +Gz exposure, blood pressure decreases at the brain level whereas increases in the lower extremities, resulting in cognitive functions impairment.

Then another study (Slungaard et al., 2017), last carried out by the UK Royal Air Force (RAF) in 2005, investigated G-induced loss of consciousness (G-LOC) within their military aircrew, leading to the introduction of subsequent interventions.

The study concluded that the incidence of G-LOC and almost loss of consciousness (A-LOC) events in aircrew could be significantly reduced through various measures. These included participation in formalized G theory classroom lectures, proper execution of the Anti-G Straining Maneuver (AGSM), and undergoing centrifuge training.

In a study of a cross-section of aerobatic pilots, Ercan and Gunduz (2020) determined that withstanding the negative effects of G forces is positively correlated with pilots' readiness and countermeasures that were taken against them. Therefore, factors that decrease resistance to G forces will increase G-induced cognitive impairment. Changes in blood pressure affect cerebral perfusion and disrupt high cortical functions. The authors also suggested that cognitive functions are less affected in older aged and more experienced pilots. Height was considered to influence G tolerance but provided small contributions. They concluded that cardiovascular changes were responsible for the cognitive dysfunctions caused by +Gz exposure. Increased pilots' ability to recognize acceleration forces and know how to cope with their results in increased adaptation to G forces over the years.

One tragic example that highlights the importance of understanding these effects is the 2015 Shoreham Airshow accident. A vintage Hawker Hunter aircraft crashed into a busy roadway, causing 11 fatalities and several injuries (Air Accidents Investigation Branch, 2017). The investigation suggested that the pilot had been subjected to high G forces and may have experienced cognitive impairment or even temporary loss of consciousness, which could have contributed to the accident. This tragic event underlines the need for comprehensive research and awareness of the physiological and psychological factors affecting air show performers.

4.5.1 G Forces

G-induced vestibular dysfunction (GIVD) is a condition that affects air show performers and competition aerobatic pilots, often occurring during high negative G maneuvers (less than negative four Gs) or when pilots experience a high positive G load following a negative G load, also known as the push-pull effect (Muller, 2002). GIVD can cause symptoms such as vertigo, an unstable gait, and nausea without vomiting, commonly referred to as "wobblies." Even a healthy vestibular system can be negatively impacted within the aerobatic environment, and some vestibular disorders, such as GIVD, may become exaggerated or impaired during aviation activities (Demir & Aydin, 2021).

Although there is no clear evidence directly linking negative Gs to consequences in cognitive performance for air show performers, it is still worth considering the potential for discomfort and other effects similar to those caused by positive Gs. Positive Gs are known to cause G-induced loss of consciousness (G-LOC) and other physiological effects, such as reduced blood flow to the brain, affecting cognitive performance, decision-making, and emotional state (Green, 2016). While negative Gs are different in nature, they still subject the body to rapid changes in blood flow and pressure, which could potentially have an impact on psychological factors.

Negative Gs cause a blood flow shift toward the head, resulting in discomfort and potentially red outs, which could reduce visual acuity due to excessive blood flow to the head, which can also affect visual sensory modality, perceptual encoding, and decision-making (Keebler et al., 2022). If negative G forces exceed about two or three

Gs, then the increased blood in the head can cause burst blood vessels. This commonly happens in either the eyes or the brain as the vessels are overloaded by the extra blood being forced through them. This can lead to permanent blindness or brain damage.

Although there is a lack of specific research on the psychological effects of negative Gs on air show performers, the physiological discomfort they experience could potentially affect their reaction time and emotional state in a manner similar to positive Gs. To accurately assess the psychological effects of negative Gs on air show performers, further research is needed. It is essential to examine the unique challenges faced by these pilots and explore the potential impact of negative Gs on their psychological well-being.

4.5.2 Push-Pull Effect

The push-pull effect experienced by air show performers involves a rapid transition between high negative G forces and high positive G forces during flight maneuvers. This sudden change can have significant physiological consequences, leading to discomfort and potential complications. Although direct evidence linking the push-pull effect to psychological factors is limited, understanding its potential impact on air show performers' emotional state, decision-making, and cognitive performance is crucial.

There have been some research-based studies linking the push-pull effect to physiological factors such as transient incapacitation and almost loss of consciousness, which can have detrimental effects on a performer's cognitive function, such as attentional resources, decision-making, and reaction time (Banks et al., 2011; Ercan & Gunduz, 2020). The physiological discomfort resulting from the push-pull effect may trigger heightened anxiety, stress, and other negative emotions. These emotional responses can impair a pilot's ability to concentrate and maintain focus during demanding aerial displays. As a result, air show performers may be more prone to making errors or engaging in risky behaviors due to the emotional strain associated with the push-pull effect.

The push-pull effect can potentially influence a pilot's decision-making capabilities by increasing cognitive load and compromising their ability to process information efficiently. The rapid transition between positive and negative Gs may lead to a loss of spatial awareness and confusion. Such transitions could adversely affect the attentional process, information retrieval, decision-making, and eventually reaction time, especially under such a high-stress environment. Some air show performers may struggle to execute complex maneuvers and maintain situational awareness during such transitions. An impaired decision-making process can increase the risk of errors resulting in accidents which can jeopardize the safety of both pilots and spectators.

To mitigate the harmful effects of the push-pull effect on air show performers' psychological state, several strategies can be considered:

- Comprehensive Training

Ensure pilots receive comprehensive training that explicitly addresses the push-pull effect and its potential impact on psychological factors. This training should include techniques for managing physiological discomfort and maintaining focus during high-stress situations.

- Regular Practice

Encourage air show performers to practice and rehearse their routines consistently, gradually increasing the intensity and complexity of their maneuvers. This will help build their confidence, familiarity, and skill in managing the push-pull effect.

- Physical Conditioning

Encourage air show performers to maintain a high level of physical fitness, including strength, flexibility, cardiovascular endurance, and sleep. Physical conditioning can help pilots better tolerate the physiological effects of the push-pull effect, potentially reducing its impact on psychological factors.

- Gradual Exposure

Design training programs that gradually expose pilots to the push-pull effect, allowing them to acclimate to the sensations and physiological changes associated with rapid transitions between positive and negative Gs. This can help pilots build resilience and adapt to the push-pull effect more effectively.

- Effective Communication

Foster open communication among air show performers, coaches, and support staff to ensure any concerns or difficulties related to the push-pull effect are identified and addressed promptly. This can help create a supportive environment that prioritizes pilots' mental and physical well-being.

- Peer Support

Encourage air show performers to share their experiences and strategies for managing the push-pull effect with their peers. Peer support can help pilots learn from one another and develop a sense of camaraderie that may mitigate the psychological impact of the push-pull effect.

- Monitor Performance

Regularly monitor air show performers' performance and psychological well-being to identify any signs of stress, anxiety, or other issues related to the push-pull effect. Early identification of potential problems can enable timely interventions and support.

4.5.3 Low-Level Aerobatics

Low-level aerobatics presents unique challenges for air show performers, as they require increased situational awareness, precise control, and rapid decision-making.

FIGURE 4.4 An air race pilot flying under the Chain Bridge, Budapest, in Red Bull Air Race® 2008. Copyright: Paul Williams

The high-stakes nature of flying close to the ground can heighten pilots' stress levels, potentially affecting their cognitive and emotional states.

The increased risk associated with low-level aerobatics might exacerbate the mental workload experienced by pilots. Managing altitude, airspeed, and aircraft control while maintaining situational awareness and anticipating potential hazards can be a cognitively demanding task. This heightened mental workload could impair pilots' decision-making abilities, making it difficult for them to respond effectively to changing conditions or unexpected events.

Additionally, low-level aerobatics might increase the likelihood of pilots becoming fixated on specific aspects of their performance or flight situation. This fixation can detract from their overall situational awareness and may impair their decision-making capabilities, potentially leading to dangerous outcomes.

To mitigate the psychological impact of low-level aerobatics, air show performers might consider engaging in extensive training and practice at low altitudes, allowing pilots to become more comfortable and confident in their abilities. This familiarity might help to alleviate stress and reduce the mental workload associated with low-level aerobatics.

Another potential strategy is to incorporate stress management techniques, such as mindfulness exercises and relaxation techniques, into pilots' training routines. These practices could help pilots remain calm and focused during high-pressure situations, ultimately improving their decision-making abilities.

Pilots might also benefit from developing contingency plans and rehearsing emergency procedures for various scenarios, including those specific to low-level aerobatics. By being prepared for unexpected events, pilots might be better equipped to make rapid and effective decisions in response to unforeseen circumstances.

In summary, low-level aerobatics presents unique psychological challenges for air show performers, with potential impacts on decision-making, mental workload, and performance. Engaging in realistic or scenario-based training, stress management techniques, and contingency planning could enable pilots to allocate more cognitive resources in the handling of the high workload and emotional demands of low-level aerobatics (see Figure 4.4).

4.6 CASE STUDY: THE 2015 SHOREHAM AIRSHOW TRAGEDY: THE ROLE OF PSYCHOLOGICAL FACTORS IN THE PILOT'S DECISION-MAKING

4.6.1 Introduction

The 2015 Shoreham Airshow accident, in which a Hawker Hunter aircraft crashed into the A27 road, killing 11 people, and injuring 13 others, including the pilot, highlights the importance of understanding the psychological factors that may influence pilots' decision-making during air shows (Air Accidents Investigation Branch, 2017). This case study explores the psychological aspects that may have played a role in the pilot's decision-making and situational awareness during the accident.

4.6.2 Background

On August 22, 2015, the Hawker Hunter aircraft was performing a maneuver involving both pitching and rolling components when it failed to achieve sufficient altitude to complete the maneuver safely. The pilot did not execute an escape maneuver, and the aircraft ultimately crashed into the ground. The investigation revealed several factors that may have contributed to the pilot's failure to recognize the need for an escape maneuver, including issues related to workload, distraction, visual limitations, and misreading or incorrectly recalling the minimum height required (AAIB, 2017).

4.6.3 Psychological Factors Affecting the Pilot's Decision-Making

The investigation by the Air Accidents Investigation Branch (AAIB) examined the circumstances surrounding the crash of the vintage Hawker Hunter G-BXFI, and it revealed a number of psychological factors that contributed to the accident. Below we will analyze the main factors that could have affected the pilot's decision-making process and conclude with the implementation of the *5B* model as a framework for further analysis.

- Workload and Cognitive Capacity

The pilot's failure to recognize the need for an escape maneuver may have been influenced by the high cognitive demands placed on him during the performance. The combination of low entry speed and low engine thrust in the upward half of the maneuver may have increased the pilot's workload, making it more difficult for him to accurately perceive the situation and make appropriate decisions (AAIB, 2017).

- Distraction and Situational Awareness

The pilot may have been distracted by other tasks or aspects of the performance, reducing his ability to maintain situational awareness. Distractions can lead to critical information, such as altitude, being overlooked or misinterpreted, which can have severe consequences in high-stakes situations like air shows.

- Visual Limitations and Misinterpretation of Information

The investigation found that defects in the altimeter system may have led to the pilot receiving incorrect height information at the apex of the maneuver (AAIB, 2017). Additionally, visual limitations such as contrast or glare could have made it difficult for the pilot to accurately read the altimeter, further compromising his situational awareness.

- Training and Assessment

The pilot's lack of formal training and assessment in escape maneuvers for the Hunter aircraft may have limited his ability to recognize the need for an escape maneuver and execute it successfully (AAIB, 2017). Inadequate training can lead to gaps in pilots' understanding of critical procedures and limit their ability to adapt to changing circumstances during a performance.

4.6.4 5B Model Analysis

To implement an innovative approach to air show safety, it was deemed crucial to analyze the biases related to the Shoreham Airshow accident using the *5B* model discussed earlier in this chapter. The *5B* model provides a structured framework to understand the behavioral and psychological factors that can influence decision-making in complex and high-pressure situations. By examining the accident through this lens, it becomes possible to identify potential biases that may have contributed to the tragic outcome.

4.6.4.1 Bias 1. Plan Continuation

One of the primary psychological factors was continuation bias, which is the tendency to persist in the course of action despite evidence suggesting that it may be time to change direction. In this case, the pilot continued with the maneuver despite having flown too low and too slow to complete it safely. The desire to complete the aerobatic maneuver by the mishap pilot, notwithstanding the fact that all the flight parameters required a wave off could have been influenced by the fact that decision-making in such a rapidly changing and dynamic environment is "frontloaded" and involves assessing and reassessing it for continuous do ability. Also, strong initial and current cues might have suggested that the situation was under control and could be continued without risk (Dekker, 2014).

4.6.4.2 Bias 2. Overconfidence and Illusory Superiority

Another factor was overconfidence bias, which refers to the tendency to overestimate one's abilities and underestimate risks. The pilot had extensive experience in flying the Hawker Hunter, and this may have led him to believe that he was capable of completing the maneuver safely. However, he had not received formal training on how to escape from the maneuver in a Hunter, and his competence had not been assessed.

4.6.4.3 Bias 3. Confirmation

Confirmation bias was also evident in this case, which is the tendency to seek out information that supports one's pre-existing beliefs and to ignore evidence to the contrary. The minimum height required for the apex of the maneuver was unclear, and there was confusion among pilots about what that height should be. It is possible that the pilot relied on his previous experience and assumed that the height was higher than it actually was.

4.6.4.4 Bias 4. Anchoring

Anchoring bias may have also contributed to the accident, as the pilot relied on the perception that he had sufficient energy to complete the loop. It seems that he may become anchored to his main display profile, leaving no room to adapt to an escape maneuver when the flight parameters were not favorable to complete his maneuver safely.

4.6.4.5 Bias 5. Social Facilitation

Social facilitation bias may have played a role as well, which is the tendency to perform better in front of an audience. The Shoreham Airshow was a high-profile event with a large audience, and this may have put pressure on the pilot to perform well. Additionally, the lack of formal safety management systems in place may have led to a sense of complacency among those involved in the planning and execution of the event.

In hindsight, a critical look at the Shoreham Airshow accident using the *5B* model puts into perspective the potential effects of biases on an air show pilot's performance. These biases can be prevalent in high-pressure situations and challenging to recognize and overcome. However, by raising awareness of the safety risk posed by these biases and implementing strategies to mitigate their effects, it may be possible to improve the safety of air show performances.

4.6.5 Lessons Learned

The 2015 Shoreham Airshow accident underscores the importance of addressing the psychological factors and related biases that can influence pilots' decision-making and situational awareness during air shows. To improve safety, the following lessons can be learned from this tragic event:

- Enhance Training and Assessment Programs

Air show performers should receive comprehensive training and assessment in critical procedures, such as escape maneuvers, to ensure they are fully prepared for high-stress situations like air shows. This could include the use of high-fidelity simulators or VR to help re-enact such scenarios as part of recurrent or periodic training in recognizing the need for an escape maneuver and effectively executing it. In addition, the introduction of single-pilot resource management (SRM) and, in some instances, crew resource management, as well as threat error management (TEM) modules, could be adapted to the air show industry and be beneficial to air show performers.

- Improve Situational Awareness (SA)

Training programs should emphasize the importance of maintaining situational awareness during high-stress situations and complex maneuvers. This includes strategies for managing workload, minimizing distractions, and accurately interpreting critical information.

Furthermore, the advent of new technologies could improve SA and, ultimately, safety for air show performers. Altitude and upset recovery signals on primary flight displays are examples of such technology, as is adaptive automation, which automatically recovers an aircraft and pulls it away from the ground when specific energy management criteria are exceeded, and the pilot does not react.

- Address Visual Limitations and Information Misinterpretation

Pilots should be trained to recognize and address potential visual limitations and information misinterpretation issues that may arise during performances, such as contrast or glare affecting the readability of instruments.

- Implement Safety Management Systems

The investigation revealed that the parties involved in the planning, conduct, and regulatory oversight of the flying display did not have formal safety management systems in place to identify and manage hazards and risks (AAIB, 2017). Implementing safety management systems can help to clarify responsibilities, identify potential hazards, and establish risk mitigation strategies for air shows and other high-stakes aviation events.

- Enhance Communication and Collaboration

Improved communication and collaboration among air show organizers, regulators, and pilots can help to ensure that all parties understand their respective roles and responsibilities in managing risk and ensuring safety. This includes clearly defining responsibilities for assessing and managing hazards associated with the aircraft's display and the safety of the public outside the display site.

- Review and Update Regulations and Guidance

The investigation found that the guidance concerning the minimum height at which aerobatic maneuvers may be commenced was not applied consistently and may be unclear (AAIB, 2017). Reviewing and updating regulations and guidance related to air show performances can help to ensure that they are clear, consistent, and effective in promoting safety.

4.6.6 Conclusion

The 2015 Shoreham Airshow accident serves as a tragic reminder of the importance of understanding and addressing the psychological factors that can influence pilots' decision-making and situational awareness during high-stress situations like low-level aerobatic displays. By learning from this event and implementing changes to training, assessment, safety management, and regulatory oversight, it is possible to reduce the risk of similar accidents in the future and promote the safety of air show performers and spectators alike.

NOTES

1. The Dunning-Kruger effect is a cognitive bias, first identified in a 1999 study by social psychologists David Dunning and Justin Kruger. The effect refers to a cognitive bias in which individuals with low ability in a specific domain tend to overestimate their competence, while those with higher ability underestimate their competence. This phenomenon occurs because people with low ability often lack the metacognitive skills needed to accurately assess their own performance. The Dunning-Kruger effect emphasizes the importance of self-awareness, feedback, and continuous self-improvement to avoid overconfidence or underconfidence.
2. The "60-minute rule" is designed to minimize distractions and maximize focus for air show performers. According to this rule, performers should avoid interaction with the crowd for at least 60 minutes before their scheduled display. This helps to ensure that performers can concentrate fully on their routines, perform necessary pre-flight checks, and mentally prepare for their performance.

LIST OF REFERENCES

Air Accidents Investigation Branch. (2017). *Report on the accident to Hawker Hunter T7, G-BXFI near Shoreham airport on August 22 2015* (Aircraft Accident Report No. 1/2017; p. 452). Department of Transport, United Kingdom. https://assets.publishing.service.gov.uk/media/58b9247740f0b67ec80000fc/AAR_1-2017_G-BXFI.pdf

Banks, R. D., Brinkley, J. W., Allnutt, R., & Harding, R. M. (2011). *Human response to acceleration* (pp. 83–109).

Barker, D. (2003). *Zero error margin: Airshow display flying analysed.* https://www.researchgate.net/publication/261136368_Zero_Error_Margin_-_Airshow_Display_Flying_Analysed

Barker, D. (2020). *Anatomy of airshow accidents* (1st ed., Vol. 1). International Council of Air Shows, International Air Show Safety Team.

Dekker, S. (2014). *Safety differently: Human factors for a new era* (2nd ed.). CRC Press.

Dekker, S. (2017). *The field guide to understanding "human error"* (3rd ed.). CRC Press. https://www.taylorfrancis.com/books/mono/10.1201/9781317031833/field-guide-understanding-human-error-sidney-dekker

Demir, A., & Aydin, E. (2021). Vestibular illusions and alterations in aerospace environment. *Turkish Archives of Otorhinolaryngology*, *59*, 139–149. https://doi.org/10.4274/tao.2021.2021-3-3

Diehl, A. E. (2002). *Silent knights: Blowing the whistle on military accidents*. Brassey's Inc.

Dunning, D. (2011). The Dunning–Kruger effect: On being ignorant of one's own ignorance. In J. M. Olson & M. P. Zanna (Eds.), *Advances in experimental social psychology* (Vol. 44, pp. 247–296). Academic Press. https://doi.org/10.1016/B978-0-12-385522-0.00005-6

Ercan, E., & Gunduz, S. H. (2020). The effects of acceleration forces on cognitive functions. *Microgravity Science and Technology*, *32*(4), 681–686. https://doi.org/10.1007/s12217-020-09793-0

Federal Aviation Administration. (2016). *Pilot's handbook of aeronautical knowledge* (2016th ed.). Federal Aviation Administration. https://www.faa.gov/regulations_policies/handbooks_manuals/aviation/phak/media/pilot_handbook.pdf

Fiske, S. T., & Taylor, S. E. (2021). *Social cognition: From brains to culture* (4th ed.). Sage Publications. https://study.sagepub.com/fiskeandtaylor3e

Green, N. D. C. (2016). Long duration acceleration. In *Ernsting's aviation and space medicine 5E* (5th ed.). CRC Press, Boca Raton, FL.

International Council of Air Shows. (2012, October 1). Sacred sixty minutes: An update from the field. *ICAS Operations Bulletin*, *5*(11). https://airshows.aero/GetDoc/2924

Keebler, J. R., Lazzara, E. H., Wilson, K. A., & Blickensderfer, E. L. (Eds.). (2022). *Human factors in aviation and aerospace* (3rd ed.). Academic Press. https://doi.org/10.1016/B978-0-12-420139-2.00004-6

Kern, M. T. (1995). *Darker shades of blue: A case study of failed leadership*. 27.

Kruger, J., & Dunning, D. (1999). Unskilled and unaware of it: How difficulties in recognizing one's own incompetence lead to inflated self-assessments. *Journal of Personality and Social Psychology*, *77*(6), 1121–1134. https://doi.org/10.1037//0022-3514.77.6.1121

Martinussen, M., & Hunter, D. R. (2018). *Aviation psychology and human factors* (2nd ed.). CRC Press.

Muller, T. U. (2002). G-induced vestibular dysfunction ('the wobblies') among aerobatic pilots: A case report and review. *Ear, Nose, & Throat Journal*, *81*(4), 269–272.

Papadakis, M. (2008). *An initial study to discover the human factors underlying air display accidents* [Unpublished master's thesis]. Cranfield University.

Rafaeli, S., & Noy, A. (2002). Online auctions, messaging, communication and social facilitation: A simulation and experimental evidence. *European Journal of Information Systems*, *11*(3), 196–207. https://doi.org/10.1057/palgrave.ejis.3000434

Slungaard, E., McLeod, J., Green, N. D. C., Kiran, A., Newham, D. J., & Harridge, S. D. R. (2017). Incidence of G-induced loss of consciousness and almost loss of consciousness in the Royal Air Force. *Aerospace Medicine and Human Performance*, *88*(6), 550–555. https://doi.org/10.3357/AMHP.4752.2017

Thompson, S. C. (1999). Illusions of control: How we overestimate our personal influence. *Current Directions in Psychological Science*, *8*, 187–190. https://doi.org/10.1111/1467-8721.00044

Tversky, A., & Kahneman, D. (1974). Judgment under uncertainty: Heuristics and biases. *Science (New York, N.Y.)*, *185*(4157), 1124–1131. https://doi.org/10.1126/science.185.4157.1124

U.S. Air Force. (1994). *Summary of AFR 110-14 USAF accident investigation board report* (Aircraft Accident Investigation Report No. 110–14). United States Air Force. https://www.baaa-acro.com/crash/crash-boeing-b-52h-stratofortress-fairchild-afb-4-killed

Wickens, C. D., Gutzwiller, R. S., & Santamaria, A. (2015). Discrete task switching in overload: A meta-analyses and a model. *International Journal of Human-Computer Studies*, *79*, 79–84. https://doi.org/10.1016/j.ijhcs.2015.01.002

FURTHER READING

Bles, W. (1998). Coriolis illusion. In G. R. F. E. Harris & M. R. Jenkin (Eds.), *Vision and action* (pp. 163–174). Cambridge University Press.

Caldwell, J. A., Mallis, M. M., Caldwell, J. L., Paul, M. A., Miller, J. C., & Neri, D. F. (2009). Fatigue countermeasures in aviation. *Aviation, Space, and Environmental Medicine*, *80*(1), 29–59.

Dekker, S. (2011). *Drift into failure: From hunting broken components to understanding complex systems*. CRC Press/Balkema.

Harris, D. (2016). *Human performance on the flight deck*. CRC Press/Taylor & Francis Group.

Jones, D. R., O'Connor, M. E., & Boll, P. A. (2011). The effects of negative Gz on pilot performance. *Aviation, Space, and Environmental Medicine*, *82*(4), 397–406.

Kahneman, D., Sibony, O., & Sunstein, C. R. (2021). *Noise: A flaw in human judgment*. Little, Brown Spark.

Lyons, T. J., Ercoline, W., O'Toole, K., & Grayson, K. (1999). The effects of motion sickness and sopite syndrome on the performance of aviators during an instrument landing system approach. *Aviation, Space, and Environmental Medicine*, *70*(9), 851–857.

Moore, D. A., & Healy, P. J. (2008). The trouble with overconfidence. *Psychological Review*, *115*(2), 502–517.

Nickerson, R. S. (1998). Confirmation bias: A ubiquitous phenomenon in many guises. *Review of General Psychology*, *2*(2), 175–220.

Orasanu, J., & Martin, L. (1998). Errors in aviation decision making: Bad decisions or bad luck? *The International Journal of Aviation Psychology*, *8*(2), 155–177.

Shappell, S., & Wiegmann, D. A. (2000). *The human factors analysis and classification system–HFACS*. Office of Aviation Medicine, Federal Aviation Administration.

Smith, D. G. (2018). Identifying and managing stress in aerobatic pilots. *Journal of Aviation/Aerospace Education & Research*, *27*(2), 53–63.

Staal, M. A. (2004). *Stress, cognition, and human performance: A literature review and conceptual framework*. NASA Ames Research Center.

Turner, M., & Griffin, M. J. (2004). Motion sickness in public road transport: Passenger behavior and susceptibility. *Ergonomics*, *47*(3), 329–341.

Wickens, C. D., Hutchins, S. D., Laux, L., & Sebok, A. (2015). The impact of sleep disruption on complex cognitive tasks: A meta-analysis. *Human Factors*, *57*(6), 930–946.

Zajonc, R. B. (1965). Social facilitation. *Science*, *149*(3681), 269–274.

5 Mindfulness

5.1 INTRODUCTION

In the world of air show performances, where precision, skill, and mental fortitude are essential, pilots often face immense pressure to perform flawlessly, which can strain their mental capacity. We have previously emphasized the significance of mindfulness as a means to counter biases or factors that interfere with the performers' mental processes during their performances. However, in this chapter, we comprehensively explore mindfulness as a tool to mitigate the potential adverse effects of psychological factors affecting air show performers.

While mindfulness practices are already established within the air show community, it is crucial to recognize when these practices have been employed and how we can further incorporate mindfulness as an additional resource in our safety toolbox. We also delve into the benefits of mindfulness for air show performers and offer practical guidance on incorporating this practice into their daily routines. We also examine two case studies: the French Air and Space Force's Rafale Solo Display and the U.S. Navy Demonstration Team, the Blue Angels. We sum up effective mindful strategies to ensure optimized performance and safety in the air show industry.

5.2 MINDFULNESS

The existing literature contains a plethora of studies on the concept of mindfulness (Baltzell & Akhtar, 2014; Birrer et al., 2012; Cole et al., 2015; Gethin, 2011; Holas & Jankowski, 2013; Kabat-Zinn, 1994; Katz et al., 1956; Krieger, 2005; Li et al., 2020; Nilsson & Kazemi, 2016; Shonin & Van Gordon, 2015; Stocker et al., 2017), yet, its exact definition remains vague. In an extensive study, Nilsson and Kazemi (2016) identified 33 definitions of mindfulness and five core elements: attention/awareness, external events, ethical mindedness, cultivation, and present-centeredness.

One of the most cited historical definitions of *mindfulness* is Kabat-Zin's (1994, p. 4): "Paying attention in a particular way: On purpose, in the present moment, and nonjudgmentally." Wallace (2006, p. 59) expanded on mindfulness by explaining that it is a state in which distraction and forgetfulness do not exist. Additionally, Brown, Ryan, and Creswell (2007, p. 212) defined mindfulness as "a receptive attention to and awareness of present events and experience."

DOI: 10.1201/9781003431879-5

While the concept of mindfulness is based on Asian spiritual/philosophical frameworks, it has almost nothing to do with religion (Kabat-Zinn, 1994). According to Kabat-Zin, mindfulness should not conflict with one's culture, traditions, or beliefs; instead, it can fill gaps in the process of one's self-development.

5.2.1 Individual Mindfulness

Mindfulness can be individual and collective (Reason, 2016). Individual mindfulness leads to systemic resilience, and collective mindfulness needs organizational support to improve the pilots' foresight and error wisdom. According to Reason, an individual's mental preparedness or mindfulness is more important than the technical skills required to achieve excellence in a task.

Experts in the domains of aviation psychology have begun to recognize the potential benefits of mindfulness. It has been suggested that mindfulness can improve the levels of safety performance by supporting pilots in managing job-related anxiety and feelings of burnout more effectively (Li et al., 2020). Mindfulness has been suggested as a method to optimize airline pilots' safety behavior (Ji et al., 2018). Modern technology, which has generally made aviation safer, could paradoxically compromise individual mindfulness in certain situations. This is because aircraft operators may not be as prepared or experienced in handling uncommon and severe incidents, as the technology tends to prevent such situations from occurring frequently. Thus, although safer modern technology has overall improved the aviation industry, it may inadvertently reduce individual mindfulness in aircraft operators when they face rare and severe incidents, as they have fewer opportunities to experience and manage such situations (Oliver et al., 2019).

5.2.2 Mindfulness Training

The effective practice of mindfulness, although seemingly simple, requires commitment, effort, and discipline (Kabat-Zinn, 1994). Automaticity, habitual unawareness (Kabat-Zinn, 1994), and chronic distractibility (Wallace, 2006) are the main obstacles to being mindful while still paying attention. Meditation practice (Kabat-Zinn, 1994) and mindfulness training (Ricard et al., 2014) are potential methods for moderating attention.

Mindfulness training has been shown to have positive effects on multiple areas of human performance, as it enhances attention regulation, body awareness, and emotion regulation while it promotes a change in perspective, according to Holzel et al. (2011). Denkova, Zanesco, Rogers, and Jha (2020) suggested that firefighters might benefit from short-form mindfulness training to bolster their psychological resilience. Moreover, Brown et al. (2007) concluded that mindfulness has positive outcomes in several important life domains, including mental health, physical health, behavioral regulation, and interpersonal relationships.

Mindfulness training can also moderate one's ego (Cole et al., 2015; Katz et al., 1956; Shonin & Van Gordon, 2015; Stocker et al., 2017; Verney, 2009), manage and reduce stress levels (Chiesa & Serretti, 2009; Russ, 2015), and efficiently intervene in anxiety (Hofmann et al., 2010; Li et al., 2020).

Blackburn and Epel's (2018) Nobel Prize-winning research focused on the effects of meditation and mindfulness on telomeres[1]. The study revealed the astonishing ability of meditation practices to protect and maintain telomere length, which in turn contributes to increased longevity. Their findings demonstrated that factors such as chronic stress, depression, and pessimistic thinking can shorten telomeres and consequently reduce life spans. However, mindfulness and meditation can counteract these negative effects, promoting healthier telomeres and potentially extending human life. The significance of this research in the context of training is that incorporating mindfulness and meditation practices into training programs could not only improve mental well-being and cognitive performance but also promote better physical health and longevity. By including these practices, training programs may help individuals better cope with stress, regulate emotions, and maintain a more positive outlook on life, ultimately contributing to improved overall performance and well-being.

In the military, the U.S. Army developed Mindfulness-based Mind Fitness Training (MMFT, or M-fit) that assists those with post-traumatic stress disorder (PTSD) (Russ, 2015; Seppälä et al., 2014). Brintz et al. (2019) found promising results in managing chronic pain and psychosocial issues for men and women in the military after a mindfulness-based intervention.

Evidence from research demonstrates that mindfulness training could also benefit people engaged in high-performance activities. The mental skills related to failure, pressure, performance anxiety, social anxiety, and emotion control in elite athletes (Birrer et al., 2012) and circus performers (Filho et al., 2016; Ross & Shapiro, 2017), as well as their overall performance, can also be enhanced by mindfulness-based intervention.

In the professional discipline of aviation, Reason (2016) argued that providing one-off training programs is insufficient to instill the necessary mental skills in pilots. Similar to technical skills, cognitive skills need to be continually managed, practiced, and refreshed. Mindfulness training could complement existing mental training for aviation professionals engaged in high-performance activities such as aerobatics and air races (Baltzell & Akhtar, 2014; Stocker et al., 2017). Meland et al. (2015) conducted a high-performance combat aviation population study and concluded that mindfulness training is a feasible and acceptable enhancement to current mental training.

Similarly, Gautam and Mathur (2018) suggested that mindful decision makers are prone to efficient decision-making due to their openness to feedback and being less prone to misapprehending situations. Li, Chen, Xin, and Ji (2020) conducted a study on Chinese airline pilots, which concluded that an increase in the use of mindfulness strategies decreased civil pilots' anxiety during flight activities.

Mindfulness-based training has various approaches, but all share the primary goal of enhancing self-awareness. Kabat-Zinn (1994) suggested three basic exercises during mindfulness-based training: yoga, body scan, and sitting meditation. Ross and Shapiro (2017) noted that circus performers' mental preparation includes breathing techniques and imagery. To mentally prepare themselves, pilots also engage in visualization techniques, such as chair flights (where they mentally rehearse their performance while sitting in a chair), group talking rehearsals (discussing and rehearsing with fellow pilots), and walk the talk (physically walking through the planned maneuvers while verbalizing each step). These techniques, combined with the 30-minute "bubble rule,[2]" help pilots achieve a heightened state of focus and mental readiness, which is crucial for their safety and success during high-stakes aerial performances.

The end state of mindfulness is an optimal state of experience – the flow. Csikszentmihalyi (2008) defined *flow* as the experience of an entirely immersed individual in the activity they are engaged in. The psychological state of flow, developed by Mihaly Csikszentmihalyi, is a state of mind that anyone can use in life: Mountain climbers use flow to shut out their nerves about the possibility that they could be injured during their climb. Artists frequently use flow in their everyday lives to help them disconnect from the objective of finishing the piece and focus solely on the process of creating art (Csikszentmihalyi, 2008).

5.2.3 The Power of Mindfulness for Air Show Performers

Mindfulness is the practice of purposefully paying attention to the present moment without judgment (Kabat-Zinn, 1994). It involves being fully aware of one's thoughts, feelings, and bodily sensations, allowing individuals to cultivate a sense of inner harmony and mental clarity. For air show performers, mindfulness offers several key benefits, such as improved focus, stress reduction, enhanced performance, and better decision-making. Below we explore the importance of mindfulness in the context of air show performances and discuss how it can be effectively applied to improve both personal and professional outcomes.

- Improved Focus

Mindfulness has been shown to improve attention resources in various contexts (Tang et al., 2015). Air show performers are required to maintain a high level of mental concentration throughout their performances. The intricate maneuvers and high-speed aerobatics demand precise coordination and split-second decision-making. By regularly engaging in mindfulness practices, air show performers can hone their ability to remain present and focused on the task at hand, leading to increased precision and control during performances.

For example, performers can begin their day with a 10-minute mindfulness meditation, paying attention to their breath and bodily sensations and gently bringing their focus back to the breath whenever the mind wanders. This practice can help performers develop greater concentration and presence, which can be applied during training sessions and performances, ultimately resulting in increased precision and control.

- Stress Reduction

The high-stakes and adrenaline-filled nature of air show performances can be a significant source of stress for performers. Chronic stress has been linked to numerous adverse health outcomes, such as an increased risk of heart disease, depression, and anxiety (Cohen et al., 2007). Mindfulness has been demonstrated to be an effective tool for reducing stress, as it encourages individuals to acknowledge their emotions without becoming overwhelmed (Chiesa & Serretti, 2009). By incorporating mindfulness techniques into their daily routines, air show performers can better manage the stressors associated with their profession and maintain a healthier mental state.

One practical approach is to engage in brief mindful pauses during training breaks or before a performance. Performers can take a few minutes to close their eyes, take deep breaths, and focus on the sensations of their breath as it moves in and out of their bodies. This simple exercise can help performers calm their minds, reduce stress, and maintain a healthier mental state.

- Enhanced Performance and Decision-Making

The ultimate goal of any air show performer is to deliver a breathtaking and captivating performance. Research has shown that mindfulness can positively impact various aspects of performance, including decision-making, emotional regulation, and resilience (Good et al., 2016). By developing a mindfulness practice, performers can foster a greater sense of self-awareness and self-regulation, allowing them to respond to challenges and recover from setbacks more effectively. This, in turn, can lead to more consistent and captivating performances for audiences to enjoy.

To enhance performance and decision-making, performers can integrate mindfulness practices into their preperformance routines. For instance, they can engage in a body scan meditation, systematically focusing on different parts of the body, releasing tension, and promoting relaxation. This practice can help performers develop greater body awareness, enabling them to respond more effectively to challenges and recover from setbacks during performances.

Additionally, performers can practice mindfulness techniques that encourage reflection on thoughts and feelings they experienced at their displays without judgment, fostering a more balanced perspective. This can lead to better decision-making during performances, as performers can identify and prioritize the essential tasks in high-pressure situations, ultimately improving overall performance quality.

In conclusion, mindfulness offers a plethora of benefits for air show performers, from improved focus and stress reduction to enhanced performance. By incorporating mindfulness practices into their daily lives, these performers can better navigate the demands of their profession while continuing to captivate audiences worldwide.

5.2.4 Incorporating Mindfulness into Air Show Performers' Daily Routines: Practical Applications

As air show performers continue to discover the benefits of mindfulness, many are putting these techniques into practice both in the air and on the ground. We discuss practical applications of how pilots already incorporate mindfulness into their daily routines, including preflight briefings, in-flight mini-mindfulness, postflight debriefs, mindful communication, and mindfulness workshops.

- Preflight Briefings

Incorporating mindfulness into preflight briefings can help performers mentally prepare for their performances. Some air show teams have incorporated mindfulness exercises into their preflight briefings. In addition, during these briefings, performers can engage in a short mindfulness exercise, such as focusing on their breath for a few minutes, a group meditation, or guided visualization. This practice can help pilots center themselves and mentally prepare for their performance.

- In-Flight Mini-Mindfulness

Although it might not seem very practical, air show performers may still be able to practice short bursts of mindfulness, or mini-mindfulness as we prefer to call it, to maintain or regain focus. For example, they can take a few seconds to focus on their breath or tune into the sensations in their body during moments of lower intensity in their performance. This brief practice can help performers to stay concentrated on their profile. At the same time, it might assist them in resetting their mind in case of any factor that affects their display flow, ultimately leading to improved decision-making and better execution of their maneuvers.

- Postflight Debriefing

Following a performance or training session, performers can use mindfulness techniques. This might include discussing their experiences and emotions with fellow team members or writing in a journal if they are solo performers. By engaging in nonjudgmental reflection and discussing their experiences with their team, performers can identify areas for improvement and develop strategies for addressing challenges in future performances (Zeidan et al., 2010). This practice can also help performers foster better communication and collaboration within their team, ultimately improving overall performance quality.

- Mindful Communication

Mindful communication is another essential aspect of incorporating mindfulness into air show performers' daily routines. By actively listening to their teammates and being present during conversations, performers can foster stronger relationships within their team. This, in turn, can create a supportive environment that promotes resilience and better coping mechanisms when dealing with stress and setbacks.

- Mindfulness Workshops

Recognizing the value of mindfulness for air show performers, air show organizations could start offering workshops and seminars to enhance their mindfulness skills and develop a deeper understanding of its benefits. These workshops could cover topics such as meditation techniques, group discussions, and practical exercises to help performers integrate mindfulness into their daily routines. By participating in these workshops, performers can strengthen their mindfulness practice and better apply these techniques in high-pressure situations.

In summary, mindfulness offers myriad benefits to air show performers, including improved focus, stress reduction, enhanced performance, better decision-making, and increased resilience. By incorporating mindfulness techniques such as preflight briefings, in-flight mini-mindfulness, postflight debriefs, and mindfulness workshops into their daily routines, performers can better navigate the demands of their profession while continuing to captivate audiences around the world.

5.2.5 Adapting Mindfulness Techniques for High G Situations in Air Show Performances

As discussed in the previous section, mindfulness practices are frequently associated with a relaxed state, a low heart rate, and even closed eyes. However, traditional mindfulness practices may not directly apply to the unique challenges air show pilots face during high G situations. To address this, pilots must develop techniques to manage their physiological responses and maintain focus during these demanding moments. This section will explore strategies, such as breath control techniques, mental anchoring, gradual exposure training, and biofeedback training, which can help pilots maintain mindfulness during high G situations in air show performances.

- Breath Control Techniques

Breath control is crucial in managing the physiological responses associated with high G environments, including increased heart rate and blood pressure (Jerath et al., 2006). Diaphragmatic breathing, also known as deep belly breathing, can help pilots regulate their breathing patterns and promote relaxation during high G situations. By focusing on their breath and practicing controlled breathing techniques, pilots can maintain mindfulness and better manage the physical stressors they experience.

- Mental Anchoring

Mental anchoring is a powerful technique that can help pilots stay present and focused during high G maneuvers. By choosing a specific sensation, thought, or visual cue as an anchor, pilots can direct their attention to this anchor when they start to feel overwhelmed by the physical demands of a high G environment (Bishop et al., 2004). This practice can help performers stay grounded and maintain their focus, ultimately improving their performance in high-stress situations.

- Gradual Exposure Training

Gradual exposure training is a valuable method for performers to build tolerance and adapt to low-level air show performances' physical and mental stressors. By progressively increasing the intensity of their training, pilots can become more comfortable with the sensations and challenges of low-level maneuvers. As they become more familiar with these conditions, performers can better incorporate modified mindfulness techniques into their routines and maintain focus during demanding performances.

- Biofeedback Training

Biofeedback training is a method that can help pilots become more aware of their physiological responses during high G situations and develop strategies to manage them effectively (Prinsloo et al., 2013). Pilots can receive real-time feedback on their physical state during high G maneuvers by using sensors to monitor heart rate, muscle tension, and other physiological markers. This information can help performers recognize their stress responses and implement appropriate mindfulness techniques, such as breath control or mental anchoring, to maintain focus and mitigate the effects of high G forces.

In conclusion, air show pilots can adapt mindfulness techniques to manage their physiological responses effectively and maintain focus during high G situations. By incorporating breath control techniques, mental anchoring, gradual exposure training, and biofeedback training into their routines, pilots can successfully navigate the unique challenges of their profession and enhance their overall performance.

5.2.6 Maintaining Synchronized Mindfulness Among Air Show Team Members

Air show performances require precision, timing, and teamwork, with each team member executing their respective maneuvers in perfect harmony. Synchronized mindfulness, which refers to the shared state of mental focus and awareness among team members, is critical in ensuring these high-stakes performances' success and safety. This section explores the importance of maintaining synchronized mindfulness among air show team members and discusses various strategies, such as mindful communication, joint mindfulness training, preflight rituals, and the value of debriefing, that can help cultivate this shared state of awareness.

- The Importance of Synchronized Mindfulness

In air show performances, the slightest lapse in focus or miscommunication can lead to dangerous situations, putting both the performers and spectators at risk. Maintaining synchronized mindfulness among team members is essential to ensure smooth and accurate execution of complex maneuvers, as well as to promote effective communication and coordination within the team (Lamothe et al., 2016). Furthermore, synchronized mindfulness can help build trust and camaraderie among pilots, fostering a supportive environment that enhances overall performance and resilience in high-stress situations.

- Mindful Communication

Mindful communication is a critical factor in cultivating synchronized mindfulness among air show team members. By actively listening and being fully present during conversations and briefings, pilots can foster stronger relationships and develop a shared understanding of their performance objectives (Lamothe et al., 2016). Open and honest communication can help address concerns, clarify expectations, and ensure that all team members are on the same page, ultimately contributing to the development of synchronized mindfulness.

- Joint Mindfulness Training

Participating in joint mindfulness training sessions can also help air show team members cultivate synchronized mindfulness. By engaging in shared mindfulness exercises, such as group meditation or guided imagery, pilots can enhance their individual mindfulness skills while also learning to attune to the mental states of their fellow team members (Kabat-Zinn, 1994). This shared practice can foster a sense of unity and collective focus, which is critical for the successful execution of high-stakes air show performances.

- Preflight Rituals

Establishing preflight rituals is another effective strategy for promoting synchronized mindfulness among air show team members. These rituals can include group breathing exercises, visualization techniques, or brief mindfulness check-ins, which provide an opportunity for pilots to connect and synchronize their mental states before taking to the skies (Holmes & Collins, 2001). By engaging in these rituals, air show performers can create a shared sense of purpose and focus that carries over into their performance.

- Debriefing and Reflection

Postperformance debriefing and reflection sessions are essential in fostering synchronized mindfulness among air show team members. These sessions provide an opportunity for air show performers to openly discuss their experiences, share insights, and learn from one another in a nonjudgmental and supportive environment (Zeidan et al., 2010). Debriefing and reflection can help pilots identify areas for improvement, address concerns, and develop strategies for overcoming challenges in future performances. By engaging in this process, team members can enhance their individual mindfulness skills and foster a shared understanding of their performance objectives, ultimately contributing to the development of synchronized mindfulness.

In conclusion, maintaining synchronized mindfulness among air show team members is essential for ensuring high-stakes performances' success, safety, and precision. Through mindful communication, joint mindfulness training, preflight rituals, and debriefing, pilots can cultivate a shared state of mental focus and awareness that enhances their performance and fosters a supportive and resilient team dynamic.

5.3 STUDY FINDINGS

The findings from interviews and focus groups suggest that air show performers who engage in mindfulness strategies tend to exhibit lower risk tolerance and reduced hazardous attitudes. They also display enhanced risk perception and contribute to a more resilient safety culture. These results emphasize the significance of mindfulness in fostering safer performance practices and a more robust safety culture among air show performers.

Mindfulness strategies can help air show performers reduce their risk tolerance, which aligns with Meland et al. (2015), who suggested that mindfulness training effects on elite individuals in high-performance environments demonstrate a more robust safety culture and greater appreciation for lower risk levels.

Moreover, these findings corroborate prior research that shows mindfulness fosters effective decision-making (Gautam & Mathur, 2018) and has a negative correlation with pilot anxiety (Li et al., 2020). The primary methods to enhance mindfulness mentioned by the interviewees also support existing literature, indicating that air show performers mentally prepare for their performances through visualization techniques and adherence to the "30-minute bubble" rule (Barker, 2020) or the "sacred 60-minute" policy (Hollowell, 2012). However, each performer noted that they use their own personalized mindfulness techniques tailored to their needs.

Furthermore, respondents shared that they mentally prepare for their flying displays by setting go-no-go weather criteria, energy gates, and decision-making nodes for emergencies. These methods are consistent with Martinussen and Hunter's (2018, p. 305) proposal that another general approach to improving aeronautical decision-making involves creating packages of predetermined decisions for various situations encountered, including specific action triggers.

The interviews revealed several themes related to mindfulness, including visualization, exogenous factor control, preshow preparation, and consistency (refer to Figure 5.1).

5.3.1 Visualization

Air show performers utilize several mindfulness techniques through visualization, such as mission briefings, flight debriefings, chair flying, and video reviews, to stay focused before a display. These practices help them mentally prepare for their performances, ensuring they are ready to handle the challenges of the show.

Preflight briefings are essential for both team and solo pilots, as they provide an opportunity to discuss critical aspects of the performance, including obstacles, antennas, weather, and routines. The length and style of these briefings can vary significantly depending on the team's culture, ranging from short 10-minute discussions to hour-long sessions (see Figures 5.2 and 5.3).

Chair-flying techniques and walking through routines are common methods for performers to visualize their maneuvers and mentally rehearse their sequences. This helps them to be more familiar with the display lines and the sequence of maneuvers they will be executing during the show.

Analyzing flight videos captured by onboard or ground cameras is another approach to visualization, particularly for new team members. These videos can help familiarize pilots with risky activities and past incidents or accidents, emphasizing the importance of following safety precautions and learning from previous experiences.

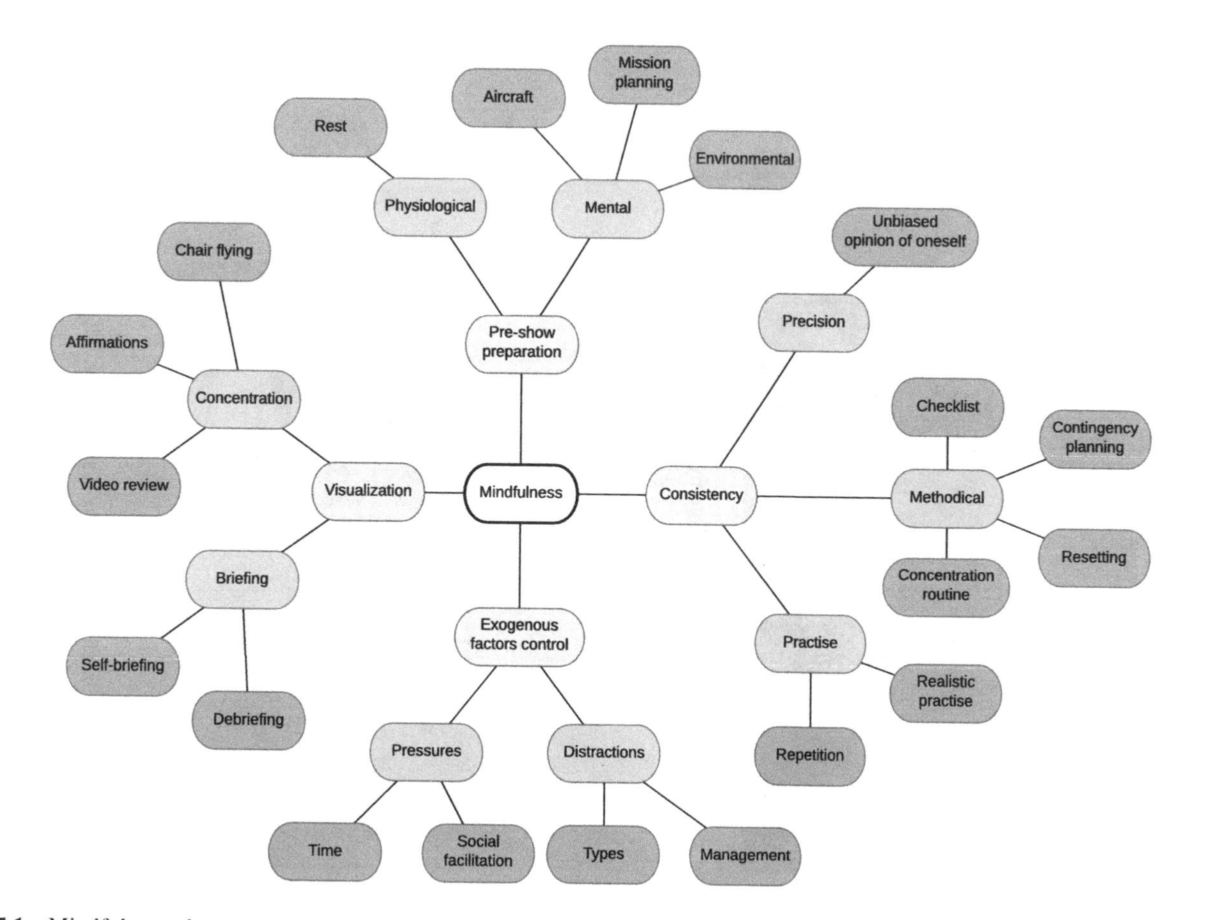

FIGURE 5.1 Mindfulness, themes map.

FIGURE 5.2 Air show safety briefing. Photo by the author (Manolis Karachalios).

FIGURE 5.3 Preflight briefing for a solo display pilot. Photo by the author (Manolis Karachalios).

Lastly, some performers incorporate affirmations as part of their mental preparation before entering the aircraft. These affirmations serve as a reminder of the inherent risks involved in air show performances and the need to remain vigilant and focused throughout the display.

5.3.2 Exogenous Factor Control

Air show performers employ various strategies to manage distractions and external pressures, such as learning from previous aerobatic competition experiences, staying focused on their tasks, seeking self-isolation, and following rules like avoiding interaction with the crowd 30–60 minutes before the display (see Figure 5.1). These strategies help performers maintain their concentration and ensure a successful performance.

One respondent highlighted the importance of having a "30-minute bubble" before flying to maintain focus. It is also essential to avoid interacting with the crowd before the display, as it can cause performers to lose concentration. Military demonstration teams often delegate the responsibility of handling public interactions to public affairs personnel to protect the performers from distractions.

Distractions are considered the second most dangerous factor in air show performances, with failure to recognize the distraction being the most dangerous. One performer mentioned that competitive aerobatics flying experience helped them improve their ability to control distractions, and they were able to transfer these skills to air show flying.

Air show performers also benefit from sterile spaces at the venue, which keep them away from crowd-induced distractions. Strict schedules are followed on air show days to manage time restrictions effectively, ensuring that performers can focus on their displays without any unnecessary pressure.

Finally, air bosses play a crucial role in providing performers with enough time buffers and minimizing constraints. They ensure that performers have adequate time between events to prepare and focus on their displays, ultimately contributing to a safer and more successful performance.

5.3.3 Preshow Preparation

Preshow preparation is a necessary mental and physiological process for air show performers that begins several days before the display (see Figure 5.1). This preparation includes planning the display, preparing footprints, mapping surroundings, and anticipating any surprises. Additionally, it involves organizing equipment and supplies at home, further reinforcing mental readiness.

According to a military interviewee, this preparation takes several days before the air display, as mentioned below:

> *The mental preparation, from my point of view, starts already several days before the air show when you prepare yourself for the display when you prepare your footprint on the surroundings, where you prepare the axes, the maps, and so on so that you think about everything, and you are not surprised by something you have not anticipated.*

Another civilian air show performer added to the above statement by noting, "It starts before an air show, it starts at home, and it starts when you prepare your stuff, get your boxing supplies ready, and your ribbon cut balls: All of that is part of the mental preparation."

Ensuring that the airplane is airworthy before the display is another essential aspect of preshow preparation. Having a ready-to-go airplane helps performers maintain focus and confidence during their performance. Respondents also emphasized the importance of mentally preparing for the current wind and weather conditions at the air show site, as these factors can impact their routine.

The wind direction can affect an air show performer's routine sequence, with some performers modifying their routines accordingly while others maintain the same sequence regardless of the wind. However, display activities such as ribbon cutting are wind dependent, making it crucial for performers to consider wind direction during their preshow preparations.

Physiological conditions are another vital aspect of preshow preparation. The majority of respondents highlighted the importance of being in good physiological shape on air show days. Ensuring adequate and proper nutrition coupled with good hydration is essential, as it contributes to optimal performance.

Lastly, securing quality sleep before an air show is another prerequisite for performers to maintain optimal physiological conditions. Adequate sleep helps ensure that performers are alert, focused, and ready to tackle the challenges of their performance, ultimately contributing to a safer and more successful display.

5.3.4 Consistency

Consistency is a critical theme in air show performers' mindful state, as it involves being methodical in conducting activities, repetition, and practice, and maintaining an unbiased opinion toward criticism (see Figure 5.1). Concentration routines and performing the same preparation sequence each time are essential aspects of consistency. Checklists also play a significant role in ensuring a thorough and consistent approach to preparation and flying.

Resetting is a safety technique used by air show performers to mitigate the risk of skipping a checklist step. By learning from their mistakes and resetting their routine, performers can reduce the likelihood of errors. Consistent and realistic practice is crucial, with performers treating practice sessions like actual air shows to maintain focus and effectiveness.

Developing consistency takes time and experience, but through repetition and the continuous pursuit of excellence, performers can stay vigilant throughout their careers. By making consistency a vital aspect of their preparation and execution, air show performers can enhance their safety and performance through strict adherence to rehearsed sequences and routines.

5.3.4.1 Live at an Air Show

During the observation period at the air show site, several mindfulness practices were identified, starting with the aircrew safety briefing in the main briefing room. These briefings were held each day of the air show and included a review of rules and regulations, weather updates, and air traffic control procedures. By examining the latest air show timeline and providing time for performers to address concerns, the aircrew safety briefing set a mindful tone for all participants and ensured everyone was prepared for their performances.

In addition to the main aircrew safety briefing, the air boss organized individual air show safety briefings as part of a fatigue risk management plan. These briefings were scheduled at different times to allow performers enough rest before their performance, particularly those performing later in the day. This approach aimed to reduce the risk of fatigue-related incidents and maintain a high level of mindfulness throughout the event.

5.3.5 Role of a Safety Observer

All military demonstration teams at the air show implemented a pivotal approach to enhance safety and mindfulness during their performances: having a safety observer stationed at the control tower. This observer maintained direct radio communication with the display pilot or team commander on a discrete frequency to minimize air traffic control-induced distractions.

Additionally, the safety observer communicated with their team's air show performers in their native language. This practice aimed to prevent any misunderstandings that could arise from language barriers when communicating with local air traffic controllers. Overall, the presence of a safety observer added an extra layer of safety and mindfulness, ensuring a smoother and more secure performance.

5.3.6 60-Minute Rule

All air show performers followed a specific practice to minimize distractions before their performances: withdrawing from social media use. Adhering to the "sacred 60-minute" rule (International Council of Air Shows, 2012), performers refrained from engaging with social media before their flights. This practice helped them maintain focus and concentration, ultimately contributing to a safer and more successful performance.

5.3.6.1 Mindfulness as Another Kit in Our Safety Toolbox

The study underscores the importance of mindfulness, which refers to being present and aware of one's thoughts and surroundings in the air show community. It reveals that mindfulness has a significant impact on risk perception, risk tolerance, and hazardous attitudes, ultimately contributing to a safer environment.

One of the highlights of the interviews was that distractions could negatively impact their ability to focus on their maneuvers and deliver a flawless performance. However, mindfulness can be a valuable tool for managing distractions and staying focused during air shows. Here are some tips for air show performers to leverage mindfulness in managing distractions:

- Recognize Distractions

The first step in managing distractions is to recognize them. Performers can stay mindful of their surroundings and notice any distractions that arise, such as the crowd or unexpected weather conditions.

- Refocus on the Performance

Once a distraction has been recognized, performers can refocus on their performance. They can use mindfulness techniques, such as deep breathing, to help them stay present and engaged at the moment.

- Let Go of Distractions

Performers can practice letting go of distractions by acknowledging them without judgment and then letting them pass. They can remind themselves that distractions are a natural part of the environment and that they can choose to refocus their attention on their performance.

- Prioritize Maneuvers

Performers can prioritize their maneuvers to help manage distractions. Focusing on the most critical maneuvers first can reduce the risk of getting distracted by less essential aspects of the show.
Mindfulness practices should be considered as an additional safety resource for everyone in the air show community, not just those at the "sharp end." Being fully present and engaged is crucial for not only performers but also air bosses and event organizers, as it helps them make more informed decisions. Ultimately, embracing mindfulness can ultimately enhance safety within the air show community.

5.4 CASE STUDIES

5.4.1 Solo Display Pilots: "Walk the Talk" Method

"Walk the talk" is a common phrase used to describe practicing what one preaches or acting following what one says. In the context of a solo display pilot, this technique could refer to a preflight mental rehearsal, visualization, or physical walkthrough of the maneuvers the pilot plans to execute during the aerial display.

This technique has several benefits for pilots:

- Familiarization

Walking through the planned maneuvers enables pilots to become more acquainted with the sequence, facilitating smoother execution during the actual performance. This familiarization process helps pilots internalize the necessary steps and timing of their routine, allowing them to operate with increased precision and confidence.

- Mental Rehearsal

Visualization of the maneuvers allows the pilot to mentally practice and reinforce the motor skills required for the aerial display. Mental rehearsal has been shown to improve performance in various sports and activities, including aviation. By mentally going through the entire performance, pilots can better anticipate challenges, refine their responses, and mentally prepare for the demands of the aerial display.

- Error Identification

Mentally rehearsing the flight helps pilots pinpoint potential errors or issues in the planned sequence, allowing for necessary adjustments before taking off. This proactive approach to error identification and correction reduces the likelihood of surprises during the performance and contributes to a safer aerial display.

- Confidence Building

Walking through the maneuvers and visualizing a successful performance can boost pilots' confidence in their abilities and preparation. Confidence is a critical factor in high-pressure situations, and feeling well-prepared can lead to improved performance and decision-making during the aerial display.

- Stress Reduction

The "walk the talk" method can also help pilots manage stress and anxiety leading up to their performance. By mentally and physically preparing for the display, pilots can gain a sense of control and mastery over their routine, which can help alleviate performance-related stress.

Many solo display pilots might use similar techniques, as mental rehearsal and visualization are well-established practices in various high-performance activities, including sports, music, and aviation. The specific details of each pilot's "walk the talk" routine may vary, but the fundamental principles of mental preparation, visualization, and rehearsal remain consistent. Technology could play a key role here with methods such as virtual reality (VR), which has been increasingly utilized for mental simulation and practice in various domains, including aviation. The integration of VR into mental simulation and rehearsal routines offers an immersive and realistic environment, simulating 3D flight profiles and allowing pilots to practice and refine their routines in a safe and controlled setting. As VR technology continues to advance, it is likely to gain wider acceptance within the aviation community, particularly among display pilots. Embracing this technology can contribute to safer and more impressive aerial displays, ultimately benefiting both the pilots and the spectators.

5.4.1.1 An Author's Perspective

The "walk the talk" method was introduced to one of the authors by his F-16 demo instructor pilot. At first, it felt unnecessary and uncomfortable, but after a few attempts, he discovered that this mental preparation significantly helped prepare him for practice or performance.

Not only did it become an ingrained routine in long-term memory, easily accessible for navigating various elements, such as adjusting for wind and weather conditions, but it also boosted his confidence. When the author left the briefing room and approached the jet, he felt prepared to perform before an audience. Both confidence and mental focus were essential. Once he closed the door and began my "walk the talk" routine, nothing was allowed to shake his confidence as he proceeded to the jet, readied himself, and followed with his display profile. The author's instructor also told him to perform the "walk the talk" behind the F-16's engine nozzles. There, behind the jet, ground crew, and onlookers, he would move his hands, pace, and recite numbers, all while dressed and ready to board the aircraft (see Figure 5.4).

While this process covers the normal flow and some specific abnormal situations, it is impossible to mentally rehearse every possible abnormal scenario. However, training, experience, and prior preparation contribute to developing the decision-making process and mental capacity needed for a safe, precise, and sharp display.

It is intriguing to note that display pilots have different preferences for their mental preparation. Some may choose a secluded room to avoid distractions, while others opt to prepare in front of their jet, connecting with the aircraft and the audience, such as the French Air and Space Force's Rafale Solo Display pilots (Karachalios, personal communication, September 24, 2022). This choice might be driven by a desire to engage with the crowd, assess the weather and environment, or be part of the ground show in front of the spectators and photographers.

FIGURE 5.4 Predisplay preparation: "Walk the talk" behind the F-16. Photo by the author (Manolis Karachalios).

5.4.2 Demo Teams: Blue Angels

The Blue Angels are the U.S. Navy's flight demonstration squadron, recognized for their precise and spectacular aerobatic displays. Several pieces of footage are available online showing the Team briefing. In the most recent one, posted in 2021, we observe the preflight briefing room at NAS Pensacola on July 7, 2021, where the pilots are preparing for a practice flight. During this process, they use mindfulness techniques to help them focus, connect as a team, and mentally prepare for the demanding tasks ahead.

Various mindfulness techniques can be seen employed by the Blue Angels during their preflight briefings, contributing to their exceptional performance:

- Visualization

The pilots "chair fly" during the brief, which involves them sitting in chairs and simulating the maneuvers they will perform in the air. By visualizing each movement in detail, they can mentally rehearse the sequence of events and reinforce their muscle memory. This process enhances their focus, coordination, and precision.

- Deep Breathing

The pilots can be seen taking slow, deep breaths as they chair fly. By focusing on their breath, pilots can improve their concentration and manage any anxiety or stress they may be feeling before the flight. Deep breathing exercises help to calm the nervous system and promote relaxation.

- Mindful Communication

During the briefing, the pilots maintain open and attentive communication with each other and their commanding officer. This allows them to stay present, listen actively, and ensure that they fully understand their roles and responsibilities.

- Mental Checklists

The Blue Angels utilize mental checklists to review critical aspects of their performance, such as safety procedures, emergency protocols, and the order of their aerial maneuvers. By mentally reviewing these checklists, they can ensure that they are fully prepared and focused on each aspect of the flight.

- Team Cohesion

The Blue Angels foster a strong sense of team unity and trust through mindfulness practices emphasizing communication, collaboration, and mutual support. By working together and supporting each other, they can perform at their highest level and mitigate risks.

These mindfulness techniques enable the Blue Angels to maintain peak mental and physical performance, ensuring they can execute their complex and hazardous maneuvers with accuracy and confidence.

Although specific information regarding mindfulness techniques used by other air show performers might not be publicly available, it is reasonable to assume that many pilots and teams within the air show community adopt similar strategies to the Blue Angels. These practices ensure safety, enhance performance and manage the stress and risks associated with their demanding profession.

NOTES

1. Telomeres are protective caps at the ends of chromosomes that play a crucial role in maintaining genomic stability. They naturally shorten as cells divide, and their length is considered a biomarker of cellular aging. Shortened telomeres have been associated with various age-related diseases and a reduced life span.
2. The "bubble rule" is a mental strategy used by display pilots to help them prepare and focus before their aerial performances. Display pilots apply the "bubble rule" 30 minutes before their performance, during which they create a metaphorical "bubble" around themselves. This bubble serves as a mental barrier, helping pilots to block out distractions, external pressures, and irrelevant thoughts, allowing them to fully concentrate on the upcoming performance.

LIST OF REFERENCES

Baltzell, A., & Akhtar, V. L. (2014). Mindfulness meditation training for sport (MMTS) intervention: Impact of MMTS with Division I female athletes. *The Journal of Happiness & Well-Being*, *2*(2), 160–173.

Barker, D. (2020). *Anatomy of airshow accidents* (1st ed., Vol. 1). International Council of Air Shows, International Air Show Safety Team.

Birrer, D., Röthlin, P., & Morgan, G. (2012). Mindfulness to enhance athletic performance: Theoretical considerations and possible impact mechanisms. *Mindfulness*, *3*(3), 235–246. https://doi.org/10.1007/s12671-012-0109-2

Bishop, S. R., Lau, M., Shapiro, S., Carlson, L., Anderson, N. D., Carmody, J., Segal, Z. V., Abbey, S., Speca, M., Velting, D., & Devins, G. (2004). Mindfulness: A proposed operational definition. *Clinical Psychology: Science and Practice*, *11*, 230–241. https://doi.org/10.1093/clipsy.bph077

Blackburn, E., & Epel, E. (2018). *The telomere effect: A revolutionary approach to living younger, healthier, longer* (2nd ed.). Orion Spring.

Brintz, C., Miller, S., Rae Olmsted, K., Bartoszek, M., Cartwright, J., Kizakevich, P., Butler, M., Asefnia, N., Buben, A., & Gaylord, S. (2019). Adapting mindfulness training for military service members with chronic pain. *Military Medicine*, *185*. https://doi.org/10.1093/milmed/usz312

Brown, K., Ryan, R., & Creswell, J. (2007). Mindfulness: Theoretical foundations and evidence for its salutary effects. *Psychological Inquiry*, *18*, 211–237. https://doi.org/10.1080/10478400701598298

Chiesa, A., & Serretti, A. (2009). Mindfulness-based stress reduction for stress management in healthy people: A review and meta-analysis. *Journal of Alternative and Complementary Medicine (New York, N.Y.)*, *15*(5), 593–600. https://doi.org/10.1089/acm.2008.0495

Cohen, S., Janicki-Deverts, D., & Miller, G. E. (2007). Psychological stress and disease. *JAMA*, *298*(14), 1685–1687. https://doi.org/10.1001/jama.298.14.1685

Cole, N. N., Nonterah, C. W., Utsey, S. O., Hook, J. N., Hubbard, R. R., Opare-Henaku, A., & Fischer, N. L. (2015). Predictor and moderator effects of ego resilience and mindfulness on the relationship between academic stress and psychological well-being in a sample of Ghanaian college students. *Journal of Black Psychology*, *41*(4), 340–357. https://doi.org/10.1177/0095798414537939

Csikszentmihalyi, M. (2008). *Flow: The psychology of optimal experience* (3rd ed.). Harper Perennial.

Denkova, E., Zanesco, A. P., Rogers, S. L., & Jha, A. P. (2020). Is resilience trainable? An initial study comparing mindfulness and relaxation training in firefighters. *Psychiatry Research*, *285*, 112794. https://doi.org/10.1016/j.psychres.2020.112794

Filho, E., Aubertin, P., & Petiot, B. (2016). The making of expert performers at Cirque du Soleil and the National Circus School: A performance enhancement outlook. *Journal of Sport Psychology in Action*, *7*(2), 68–79. https://doi.org/10.1080/21520704.2016.1138266

Gautam, A., & Mathur, R. (2018). Influence of mindfulness on decision making and psychological flexibility among aircrew. *Journal of Psychosocial Research*, *13*(1), 199–207.

Gethin, R. (2011). On some definitions of mindfulness. *Contemporary Buddhism*, *12*(1), 263–279. https://doi.org/10.1080/14639947.2011.564843

Good, D. J., Lyddy, C. J., Glomb, T. M., Bono, J. E., Brown, K. W., Duffy, M. K., Baer, R. A., Brewer, J. A., & Lazar, S. W. (2016). Contemplating mindfulness at work: An integrative review. *Journal of Management*, *42*(1), 114–142. https://doi.org/10.1177/0149206315617003

Hofmann, S. G., Sawyer, A. T., Witt, A. A., & Oh, D. (2010). The effect of mindfulness-based therapy on anxiety and depression: A meta-analytic review. *Journal of Consulting and Clinical Psychology*, *78*(2), 169–183. https://doi.org/10.1037/a0018555

Holas, P., & Jankowski, T. (2013). A cognitive perspective on mindfulness. *International Journal of Psychology*, *48*(3), 232–243. https://doi.org/10.1080/00207594.2012.658056

Hollowell, D. (2012, August 1). Putting accident/incident analysis to work for our air show family. *Air Show Digest*. https://airshowdigest.aero/2012/08/01/putting-accident-incident-analysis-to-work-for-our-air-show-family/

Holmes, P. S., & Collins, D. J. (2001). The PETTLEP approach to motor imagery: A functional equivalence model for sport psychologists. *Journal of Applied Sport Psychology*, *13*, 60–83. https://doi.org/10.1080/10413200109339004

Holzel, B., Lazar, S., Gard, T., Schuman-Olivier, Z., Vago, D., & Ott, U. (2011). How does mindfulness meditation work? Proposing mechanisms of action from a conceptual and neural perspective. *Perspectives on Psychological Science*, *6*(6), 537–559. https://doi.org/10.1177/1745691611419671

International Council of Air Shows. (2012, October 1). Sacred sixty minutes: An update from the field. *ICAS Operations Bulletin*, *5*(11). https://airshows.aero/GetDoc/2924

Jerath, R., Edry, J. W., Barnes, V. A., & Jerath, V. (2006). Physiology of long pranayamic breathing: Neural respiratory elements may provide a mechanism that explains how slow deep breathing shifts the autonomic nervous system. *Medical Hypotheses*, *67*(3), 566–571. https://doi.org/10.1016/j.mehy.2006.02.042

Ji, M., Yang, C., Li, Y., Xu, Q., & He, R. (2018). The influence of trait mindfulness on incident involvement among Chinese airline pilots: The role of risk perception and flight experience. *Journal of Safety Research*, *66*, 161–168. https://doi.org/10.1016/j.jsr.2018.07.005

Kabat-Zinn, J. (1994). *Wherever you go, there you are: Mindfulness meditation in everyday life* (10th Anniversary). First Hachette Books.

Katz, D., Sarnoff, I., & McClintock, C. (1956). Ego-defense and attitude change. *Human Relations*, *9*(1), 27–45. https://doi.org/10.1177/001872675600900102

Krieger, J. L. (2005). Shared mindfulness in cockpit crisis situations: An exploratory analysis. *The Journal of Business Communication*, *42*(2), 135–167. https://doi.org/10.1177/0021943605274726

Lamothe, M., Rondeau, É., Malboeuf-Hurtubise, C., Duval, M., & Sultan, S. (2016). Outcomes of MBSR or MBSR-based interventions in health care providers: A systematic review with a focus on empathy and emotional competencies. *Complementary Therapies in Medicine*, *24*, 19–28. https://doi.org/10.1016/j.ctim.2015.11.001

Li, Y., Chen, H., Xin, X., & Ji, M. (2020). The influence of mindfulness on mental state with regard to safety among civil pilots. *Journal of Air Transport Management*, *84*, 101768. https://doi.org/10.1016/j.jairtraman.2020.101768

Martinussen, M., & Hunter, D. R. (2018). *Aviation psychology and human factors* (2nd ed.). CRC Press.

Meland, A., Fonne, V., Wagstaff, A., & Pensgaard, A. M. (2015). Mindfulness-based mental training in a high-performance combat aviation population: A one-year intervention study and two-year follow-up. *International Journal of Aviation Psychology*, *25*(1), 48–61. https://doi.org/10.1080/10508414.2015.995572

Nilsson, H., & Kazemi, A. (2016). Reconciling and thematizing definitions of mindfulness: The big five of mindfulness. *Review of General Psychology*, *20*(2), 183–193. https://doi.org/10.1037/gpr0000074

Oliver, N., Calvard, T., & Potočnik, K. (2019). Safe limits, mindful organizing and loss of control in commercial aviation. *Safety Science*, *120*, 772–780. https://doi.org/10.1016/j.ssci.2019.08.018

Prinsloo, G. E., Rauch, H. G. L., Karpul, D., & Derman, W. E. (2013). The effect of a single session of short duration heart rate variability biofeedback on EEG: A pilot study. *Applied Psychophysiology and Biofeedback*, *38*(1), 45–56. https://doi.org/10.1007/s10484-012-9207-0

Reason, J. (2016). *The human contribution*. Routledge.

Ricard, M., Lutz, A., & Davidson, R. J. (2014). Mind of the meditator. *Scientific American*, *311*(5), 38–45. https://doi.org/10.1038/scientificamerican1114-38

Ross, A., & Shapiro, J. (2017). Under the big top: An exploratory analysis of psychological factors influencing circus performers. *Performance Enhancement & Health*, *5*, 115–121. https://doi.org/10.1016/j.peh.2017.03.001

Russ, S. (2015, July 8). Mindfulness-based stress reduction finds a place in the military. *United States Army*. https://www.army.mil/article/151787/mindfulness_based_stress_reduction_finds_a_place_in_the_military

Seppälä, E. M., Nitschke, J. B., Tudorascu, D. L., Hayes, A., Goldstein, M. R., Nguyen, D. T. H., Perlman, D., & Davidson, R. J. (2014). Breathing-based meditation decreases post-traumatic stress disorder symptoms in U.S. military veterans: A randomized controlled longitudinal study. *Journal of Traumatic Stress*, *27*(4), 397–405. https://doi.org/10.1002/jts.21936

Shonin, E., & Van Gordon, W. (2015). Practical recommendations for teaching mindfulness effectively. *Mindfulness*, *6*(4), 952–955. https://doi.org/10.1007/s12671-014-0342-y

Stocker, E., Englert, C., & Seiler, R. (2017). Mindfulness and self-control in sport. *Journal of Sport & Exercise Psychology*, *39*(Suppl.), Article Suppl. https://doi.org/10.7892/boris.112471

Tang, Y.-Y., Hölzel, B. K., & Posner, M. I. (2015). The neuroscience of mindfulness meditation. *Nature Reviews. Neuroscience*, *16*(4), 213–225. https://doi.org/10.1038/nrn3916

Verney, J. (2009). Mindfulness and the adult ego state. *Transactional Analysis Journal*, *39*(3), 247–255. https://doi.org/10.1177/036215370903900308

Wallace, B. A. (2006). Resurgent attention and close attention. In *The attention revolution: Unlocking the power of the focused mind* (1st ed., pp. 43–73). Wisdom Publications.

Zeidan, F., Johnson, S. K., Diamond, B. J., David, Z., & Goolkasian, P. (2010). Mindfulness meditation improves cognition: Evidence of brief mental training. *Consciousness and Cognition*, *19*(2), 597–605. https://doi.org/10.1016/j.concog.2010.03.014

FURTHER READING

Brach, T. (2012). *True refuge: Finding peace and freedom in your own awakened heart.* Bantam Books.

Chambers, R., Gullone, E., & Allen, N. B. (2009). Mindful emotion regulation: An integrative review. *Clinical Psychology Review*, *29*(6), 560–572.

Goleman, D., & Davidson, R. J. (2017). *Altered traits: Science reveals how meditation changes your mind, brain, and body.* Avery Publishing.

Greeson, J. M., & Chin, G. (2015). Mindfulness and physical disease: A concise review. *Current Opinion in Psychology*, *5*, 120–123.

Grossman, P., Niemann, L., Schmidt, S., & Walach, H. (2004). Mindfulness-based stress reduction and health benefits: A meta-analysis. *Journal of Psychosomatic Research*, *57*(1), 35–43.

Jha, A. P., Stanley, E. A., Kiyonaga, A., Wong, L., & Gelfand, L. (2010). Examining the protective effects of mindfulness training on working memory capacity and affective experience. *Emotion*, *10*(1), 54–64.

Lutz, A., Slagter, H. A., Dunne, J. D., & Davidson, R. J. (2008). Attention regulation and monitoring in meditation. *Trends in Cognitive Sciences*, *12*(4), 163–169.

Salzberg, S. (2011). *Lovingkindness: The revolutionary art of happiness.* Shambhala Publications.

6 Epic Survivals

6.1 INTRODUCTION

In the "Epic Survivals" section of this book, we present a collection of inspiring stories that begins with an introspective look at one of the authors' near miss with a controlled flight into terrain (CFIT) event during an aerobatic display. The open acknowledgment and sharing of stories of such close encounters by the author provide an incredible tale of successful recovery from a potentially fatal scenario. The skills, knowledge, and aptitude gained from this successful recovery by the author provide real-life experiences and an understanding of air show display complexities.

Using this personal experience and that of other air show performers, we chart through stories that provide the reader with lessons on the need for operational safety risk management in the air show industry. Also, in these pages, you will not only encounter captivating accounts of resilience and courage but also delve into the psychological factors that play a crucial role in the decision-making process of air show performers. We hope to emphasize the importance of humility and openness amongst the air show performers' community through the sharing of such personal harrowing experiences.

Throughout this section, we also highlight the importance of candidly discussing personal challenges and near misses during operational activities to create learning opportunities for each other. Such expressive and sharing culture keeps our institutional memory alive, improves safety standards, and enhances desired safety practices within the air show community. We also examine the delicate balance between risk and reward, as well as the mental and emotional fortitude required to perform under pressure. Understanding these nuances significantly improves our risk assessment and decision-making capabilities, which are essential in protecting the lives of performers and spectators alike.

In "Epic Survivals," we aim to inspire readers with the power of grit, adaptability, and humility while shedding light on the complex psychological factors and risk management practices that contribute to the continued growth and success of the air show community. Moreover, "Epic Survivals" is a testament to the value of camaraderie and mutual support within the air show community. This close-knit group of professionals is bound together not only by their passion for aviation but also by their commitment to safeguarding the lives of their fellow performers and the fans who gather to marvel at their feats.

As you immerse yourself in these fascinating tales, may you find inspiration in the strength and tenacity of those who have faced seemingly impossible challenges and emerged victorious. Let their stories remind you of the importance of diligence,

DOI: 10.1201/9781003431879-6

courage, and unwavering commitment to safety in any endeavor. And may you be moved by the extraordinary resilience of the human spirit.

6.2 CASE STUDY 1. LESSONS FROM A NEAR-MISS EXPERIENCE: HELLENIC AF F-16 DEMO TEAM *ZEUS*, 2012 (MANOLIS KARACHALIOS)

I couldn't begin the incredible survival Chapter without sharing my personal experience from a few years ago. As a survivor within the air show community, I didn't make it through by mere luck. Instead, my survival was a result of a thorough risk assessment that had been factored into my display profile and permitted by the Hellenic Air Force.

On October 28, 2012, during an air display for the National Greek Parade in Thessaloniki, I experienced a self-induced distraction that almost cost me my life. While flying an F-16 demonstration, a two-second disruption from the main task of flying the airplane led me to a descent below the minimum maneuver altitude, recovering only at 300 feet above ground level (AGL).

Planning error-tolerant display profiles, maintaining extra safety margins during maneuvers, keeping a safe distance from the crowd, and ensuring sufficient vertical separation from the ground for each individual maneuver within the flight profile are all crucial elements for keeping performers alive. This incident serves as a testament to the air show community's commitment to safety; by adhering to rules and regulations, we can continue to provide thrilling yet safe entertainment for spectators.

Before delving into the specifics of the incident, I'd like to emphasize that there are two categories of air show performers: those who have experienced a remarkable survival story like mine and those who haven't shared theirs yet. In fact, all aviators can be considered survivors after each mission, as they face numerous risks and hazards on a daily basis. Despite these challenges, with calculated risk assessments, they manage to thrive and stay safe.

6.2.1 What, When, Where: The Facts

In 2010, Greece was in a serious economic crisis due to a national debt of about €300 billion ($413.6 billion) and a 120 percent of gross domestic product to debt ratio. Greece's credit rating, which is the assessment of its ability to repay its debts, was downgraded to the lowest in the eurozone, meaning it will likely be viewed as a financial black hole by foreign investors. This resulted in the country struggling to pay its bills as interest rates on existing debts soared (CNN, 2010).

Greece was in significant breach of eurozone rules on deficit management, and with the financial markets betting the country would default on its debts, this reflected badly on the credibility of the Euro currency. There are also fears that financial doubts will infect other nations at the low end of Europe's economic scale, thereby threatening the viability of the eurozone itself. By 2012, with the economy still at risk of collapsing, an exit from the E.U. by Greece (Grexit) seemed the only way for the Greek economy to survive (*The New York Times*, 2016).

In the midst of this somber economic and political period for the Greek nation, the Ministry of Defense's leadership decided to organize an air display by the Zeus F-16 flight demonstration team of the Hellenic Air Force during the National Day parade on 28 October 2012. The ceremony was to take place in Thessaloniki, the second largest city of Greece, with the President of the Hellenic Republic, His Excellency Mr. Karolos Papoulias, attending. The air display was an effort to lift up national morale since this was the first time such a display would take place in Greece. The air display was arranged to be live on the Greek National and Satellite TV.

As part of the parade, I was tasked to depart in instrument meteorological conditions (IMC), navigate to a holding area, hold below the clouds for almost half an hour, and then fly my display after the ground forces had completed their parade. The weather in the display area was overcast at four thousand feet (4,000 ft.), with light rain and visibility at 6 kilometers. I decided to press on the show with our standard weather backup plan, the low show[1], removing all vertical maneuvers (see Figure 6.1).

Midway through the display routine, during the falcon turn maneuver (which includes an 8 G-pull climbing turn, 90 degrees away from the crowd while dispensing

FIGURE 6.1 Cockpit view during a 9-G turn over Thessaloniki, 2012. Copyright: Hellenic Air Force (HAF).

flares, followed by a descending turn back to the display line at 500 feet above the sea, see Figure 6.2), when I tried to dispense the flares during the climb, nothing actually happened, and no flares went off the airplane. Immediately, I realized that during my predisplay checks, I didn't turn on one of the switches that allowed the flares to drop off the jet.

Confidently enough, during the descending part of the maneuver, I decided to set the switch to the ON position. When I reversed my turn, I looked down on the cockpit's left console to the countermeasures panel, disregarding completely that the airplane was on a steep descent of more than 20 degrees dive, with the engine in maximum afterburner and the airspeed increasing rapidly.

After almost 2 seconds, I took my eyes back to the head-up display (HUD); at the same time, a voice callout of "Pull up, pull up" was sounded in my headset, followed by a close ground proximity alert Break-X on my HUD and multifunction displays (MFD)! I realized that I was in imminent danger, and I had to recover immediately to save the crowd, myself, and the jet.

I finally managed to recover by rolling out to 70 degrees angle of the bank, pulling as hard as I could at the flight control's limiters, and I stayed away from the foul line[2]; however, I violated my target altitude of 500 feet and leveled off at 300 feet above ground level (AGL).

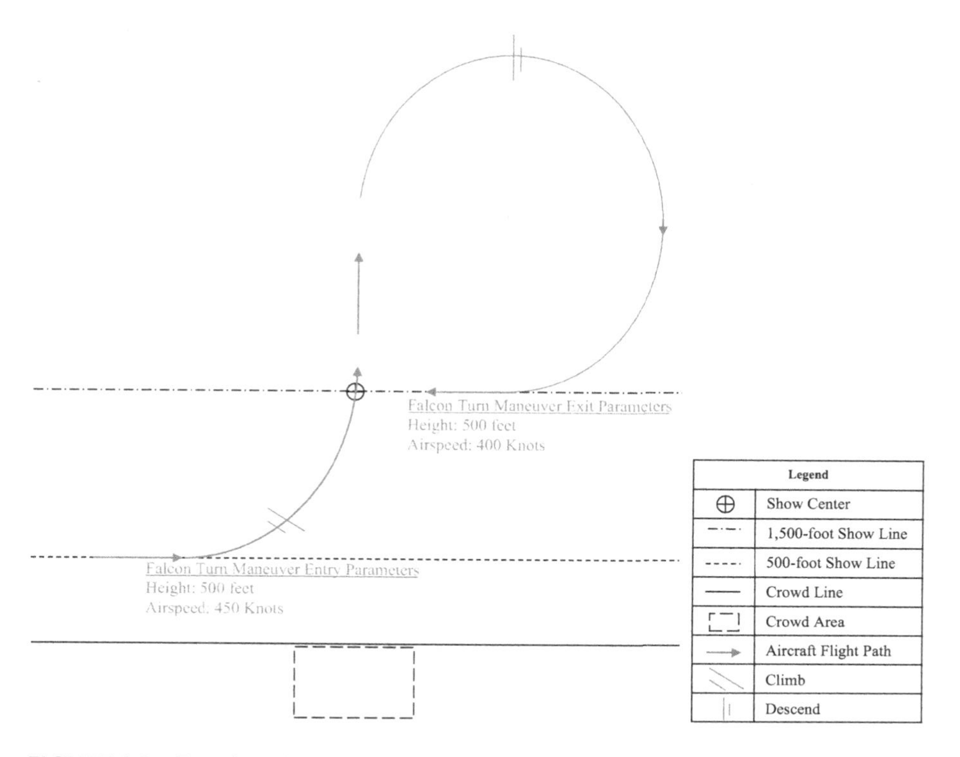

FIGURE 6.2 Top-view diagram of the falcon turn maneuver.

After the flight, when I went through the available aircraft performance documentation, it was evident that if I had stayed distracted for an extra second, the results would have been tragic.

6.2.2 Why: The Psychological Factors Involved

There might be numerous factors that contributed to the incident, yet the primary psychological factors identified include self-induced distraction, inadequate task prioritization management, social facilitation pressure, and an invulnerability attitude. Additionally, aspects of the cockpit design, such as the suboptimal placement of the countermeasures dispensing system (CMDS) rotary knob, i.e., being located outside the pilot's central field of vision, and the absence of hands on throttle and stick (HOTAS) features, played a role in the event. Using suggestions from Harris (2011, p. 14), I critically explore the human performance factors that may have been at play in this specific incident, considering, specifically, the human operator within a complex sociotechnical system.

6.2.2.1 Case Study Analysis

6.2.2.1.1 Task Switching

Although numerous factors may have contributed to the incident, this analysis focuses on task switching as a significant aspect. Task switching refers to the process of shifting between different tasks, which often incurs a time cost.

Research has shown that preparing for a switch can reduce but not eliminate this cost (Fu & Gray, 2006; C. Wickens, 2021; C. D. Wickens et al., 2013, 2015). This indicates that there are at least two components of switching cost: an intentional reconfiguration process and an external, stimulus-driven process. Notably, the costs of switching between a simple task and a more challenging one are usually higher for the simpler task.

Within the context of aviation, the flight crew's four basic tasks are aviating, navigating, communicating, and managing (Harris, 2011). When pilots follow the standard task prioritization sequence, aviating is the primary task. However, due to task switching, managing can become the primary task, replacing the critical core task of flying the aircraft. This highlights the potential risks and consequences of task switching in critical situations.

6.2.2.1.2 Central Bottleneck: Multitasking Limitations

A significant amount of research on multitasking has focused on the *psychological refractory period* (PRP) effect, which refers to the delay in reaction time for the second task when two tasks are performed in quick succession (Bock et al., 2021; Fagot & Pashler, 1992; Jung et al., 2012; C. Wickens, 2021; C. D. Wickens et al., 2015). Peripheral sensory and motor processes can contribute to decrements in dual-task performance. For example, if you are looking at the navigation display in your car, you cannot respond to visual events that occur outside, and if you are holding a mobile phone in one hand, you cannot use that hand to respond to other events.

The central bottleneck model, which is the most widely accepted explanation of the PRP effect (Held et al., 2022; Jung et al., 2012; Lien et al., 2006; Tombu & Jolicoeur, 2005), proposes that selecting a response for the second task (e.g., flying the airplane) cannot begin until the response selection for the first task (e.g., flares switch) is completed. This bottleneck imposes a limitation on multitasking performance in real-world situations, such as flying an airplane or driving a car.

6.2.2.1.3 Biases and Heuristics

Cognitive factors that influence the decision-making process, such as biases and heuristics, may have contributed to the 2012 incident, and of interest are overconfidence bias, social facilitation, plan continuation, confirmation bias, and attention tunneling. A bias is defined as "a prejudice or a propensity to make decisions while already being influenced by an underlying belief" (Chira et al., 2008, p. 12). Heuristics are mental shortcuts often used during decision-making when time, information, or resources are constrained and which typically result in a good outcome (Vidulich et al., 2010). While heuristics can be valuable in allowing display pilots to make rapid decisions and judgments, they can also introduce biases and errors. Therefore, it is essential for air show performers to be aware of these potential pitfalls and continuously train to maintain their skills, adapt their heuristics to their specific environment, and ensure that their decision-making remains as accurate and safe as possible.

We analyze the potential role of some biases and heuristics in this safety event.

1. Overconfidence Bias

Overconfidence can result in risk-taking behaviors and has been identified as a contributing factor in numerous aviation accidents (Barker, 2020b). Overconfidence bias results in an individual's tendency to overestimate the state of their knowledge and their ability to perform well (Chira et al., 2008). As a result, decision makers may reach a conclusion about the state of the situation without seeking out alternative hypotheses, which can limit decision effectiveness. Interestingly, overconfidence can be linked not only to hazardous attitudes as commonly perceived but also to a false sense of assurance stemming from an incomplete understanding of our physiological limitations. In other words, sometimes we don't know what we don't know! We might be bluntly ignorant of our human limitations.

In my experience, flying over water with a low cloud ceiling eliminated the presence of a discernible horizon and negatively impacted visual depth perception. It is important to note that I had raised my helmet's dark visor, indicating that I was experiencing visual difficulties due to the low lighting conditions resulting from the overcast weather in the display area.

When I looked down to change the switch selection, my focal vision, crucial for detecting in-flight hazards with my primary field of view, was compromised. Peripheral vision, which supplies sufficient information for display line maintenance, is relatively unaffected by such factors (Wickens et al., 2021). However, I failed to appreciate the distinct functions of each visual subsystem and assumed that adequate display line-keeping[3,4] meant that I was not impacted by low visibility conditions or cockpit distractions.

This misjudgment led to a dangerous overconfidence in my capacity to visually detect and identify inflight hazards. I was deluded in my understanding of the human sensing and perceptual systems that one visual subsystem could function safely while another was impaired.

Yet, there is another aspect of overconfidence that I experienced, one that was profoundly personal and captivating. The F-16's cockpit enveloped me in a sense of security and warmth, making me feel protected and at ease. The powerful engine roared beneath me, and the flight controls responded with precision to my every command. This connection between the aircraft and myself was so seamless that it felt as though the jet was an extension of my own body. I felt invulnerable as if nothing could harm me. This exhilarating sensation of power and control, however, was deceptive. Those are some of the defining blocks leading to overconfidence since it creates a false sense of calibration. Normally, overconfidence biases are hardly mitigated by training (Sulistyawati et al., 2011). Moreover, despite the apparent technological capabilities of the F-16 aircraft, it is essential to remember that, as humans, we remain vulnerable creatures. We must always be mindful of our limitations and strive for self-discipline, even when we feel most confident.

2. *Social Facilitation Bias*

Social facilitation bias significantly affects the decision-making process of display pilots. In my case, the pressure to perform well in front of the public during the National Parade significantly amplified the stress I experienced due to the social facilitation bias. Looking back, failure was not an option for me, as I didn't want to let down the Greek people and my family. This intense desire to succeed and meet the expectations of those around me might have further contributed to the challenges I faced during the incident.

3. *Plan Continuation Bias*

During the incident, I fell victim to plan continuation bias, commonly known in aviation circles as *get-there-itis*. This bias leads an operator such as a pilot with an original plan formed to continue with that plan, even in the face of new information that might suggest that the plan is no longer the best option (Orasanu et al., 2001). As I stubbornly adhered to my original plan, this bias instilled a sense of urgency and an overwhelming desire to complete the mission, even as the situation changed around me. In my determination, I felt compelled to drop the flares regardless of my jet's attitude – an irrational reaction in retrospect!

It's crucial to recognize when to adapt and adjust our plans to ensure safety and success in the face of evolving circumstances.

4. *Confirmation Bias*

Confirmation bias played a significant role in my decision-making process during the incident, causing me to concentrate on information that conformed to my expectations of the jet's performance. As a result, I overlooked crucial details, such as the lower apex altitude during the falcon turn maneuver, which was caused by the low cloud ceiling, resulting in a shorter time required to descend back to 500 feet.

In such situations, it's essential to remain vigilant and open-minded, continually reassessing our assumptions and actively seeking out any discrepant information. By doing so, we can avoid falling into the trap of confirmation bias and make better-informed decisions, ensuring a higher level of safety and performance in dynamic and complex environments such as display flying in challenging weather conditions.

5. Attention Tunneling

Lastly, I became excessively fixated on setting the CMDS rotary knob, causing me to lose awareness of other essential peripheral cues, such as the rapidly increasing rate of descent. This phenomenon is known as attention tunneling or, in air combat terms, "target fixation" and results in scenarios where a task receives total engagement for an extended period, to the neglect of other safety-critical tasks, such as monitoring altitude (Loukopoulos et al., 2010; C. D. Wickens & Alexander, 2009; C. D. Wickens & Yeh, 2018). Attention tunneling can impair the operator's ability to process information outside their current focus of attention.

Attentional tunneling may also have resulted in an increase in dwell duration on the CMDS rotary knob much longer than is optimal, considering its expectancy and value (Wickens & Yeh, 2018). This resulted in a long period of neglect of other high-valued areas of interest, such as a visual scan of aircraft trajectory and energy state. In the low-level aerobatic environment, I shouldn't have allowed such a high dwell time on the CMDS rotary knob. A quick glance at the switch should have been sufficient, followed by an immediate refocus on the HUD to assess the jet's flight path, energy levels, and showline repositioning.

Attention tunneling can have detrimental effects on situational awareness and decision-making, particularly in high-stress environments like low-altitude air shows. It is crucial for aerobatic pilots to maintain a balanced focus on all relevant aspects of their flight display profiles, continually scanning for potential hazards outside the cockpit (OTC) and within the cockpit (WTC). Pilots should adjust their actions accordingly without fixating on a single parameter for longer than necessary.

Through training, we can refine our "internal stopwatch" and alert ourselves if we spend too much time on a single task. If this happens, it may be an indication that something unexpected has occurred, and we should be prepared to halt our air show display profiles while resolving the issue at hand. With sound systems knowledge and experience, pilots can be able to extract vital flight information from cockpit displays in a shorter time. This approach helps ensure the safety of both the pilot and the spectators.

6.2.2.2 Conclusion

In conclusion, task switching and the central bottleneck effect significantly impact multitasking limitations, especially in high-stakes environments such as low-level display flying. The interplay of these factors can lead to prioritization issues, as seen in the analyzed incident where managing tasks took precedence over the critical task of flying my jet. Cognitive biases and heuristics, including overconfidence, social facilitation, plan continuation, confirmation bias, and attention tunneling, further exacerbated these limitations.

To address these challenges and enhance pilots' multitasking performance, it is crucial to provide comprehensive training that emphasizes the importance of proper task prioritization, especially in high-pressure situations. This training should also help pilots recognize and mitigate the influence of cognitive biases and heuristics on their decision-making processes.

Additionally, improving cockpit design by optimizing the placement of essential switches and incorporating features like hands on throttle and stick (HOTAS) functions could help reduce the need for task switching, thereby minimizing the impact of the central bottleneck effect on multitasking performance.

6.2.3 How: Risk Management Strategies that Contributed to the Epic Survival

Well-established risk management procedures in the Hellenic Air Force's Zeus Demo Team play a vital role in safeguarding air show displays, ensuring the safety of both the display pilot and spectators. These procedures encompass various aspects, such as planning display maneuvers with safety buffers and leveraging technology in the aircraft, like the ground collision avoidance system (GCAS).

6.2.3.1 Planning Display Maneuvers with Buffers

By incorporating safety buffers of energy (altitude, speed, and angle of attack available) and separation from the crowd into the design of display maneuvers, I managed to account for potential errors or unexpected factors like wind and weather conditions. This added margin of safety allowed me to adjust my performance as needed without jeopardizing the safety of myself or the crowd. More specifically, in the falcon turn maneuver, I added an extra second of pause before my reverse turn back to the crowd; this provided me with more time and space to assess the situation and make any necessary adjustments. This extra 500 feet buffer can make a significant difference when it comes to reacting to changing conditions, such as unexpected gusts of wind or a sudden decrease in the engine's available thrust.

6.2.3.2 Standard Operating Procedures

Adherence to standardized protocols and guidelines for the minimum altitudes and display lines[5] ensured that I followed best practices in terms of safety and performance. These procedures provide a structured framework for pilots, helping to reduce the likelihood of errors and enhancing the overall safety of aerial displays.

6.2.3.3 Comprehensive Training and Practice

Rigorous training, both inflight and in the simulator, played a critical role in my ability to handle the situation effectively. Typical predisplay preparation includes inflight and simulator training. By maintaining a minimum of once per week in flight training and practicing in the simulator multiple times per day, I ensured that my skills remained sharp and adaptable to various conditions.

Incorporating weather variations such as low cloud ceilings and into-crowd winds, as well as emergency scenarios, helped me to prepare for the unexpected and enhanced my ability to execute escape maneuvers or divert to alternate airfields if necessary. Additionally, the simulator training assisted me in assessing the required fuel and adjusting the fuel consumption during the show to ensure a safe return to base with sufficient fuel reserves.

Then visiting the show site on the ground with my team beforehand allowed me to gain a better understanding of the environment, such as identifying obstacles, assessing the sun's position relative to the show center, and determining any limitations in line-of-sight communication with the ground safety observer. This step was essential for planning and executing a successful and safe performance.

During the rehearsal day, a thorough briefing and practice of communication with the safety observer helped me familiarize myself with the display environment and ground features. Practicing the staged entry, display line, and holding patterns allowed me to refine my performance and anticipate any potential challenges during the actual show.

After the rehearsal flight, debriefing myself and reviewing the HUD recording and onboard camera footage provided valuable insight into the execution of individual maneuvers, timing, positioning, and flight parameters. This feedback allowed me to make necessary adjustments and learn from any mistakes or areas of improvement, ultimately leading to a more polished performance.

Research demonstrates that skill acquisition is built through the collection of multiple experiences and cases stored in long-term memory, often derived from consistent and effective practice (Anderson, 1983). These cases and examples are then recalled and acclimated to assist with the challenge at hand. This comprehensive training helped me develop automaticity in the recall of ingrained skill sets known in the air show community as "muscle memory" and situational awareness and maintain the necessary skills to recognize and respond to the arising issue. As a result, I was able to apply the appropriate technique to recover from the steep dive successfully.

6.2.3.3.1 Train as you Display, and Display as You Train

It's important to note that all my training during the rehearsal day focused solely on suitable weather conditions based on the forecasted weather the week before the display. As a result, I rehearsed only the high-show display profile. However, on the event day, a low-pressure weather system quickly approached the area of Thessaloniki, and I had to perform a low show instead. Although the low show is often considered less challenging and less spectacular than the high show for solo display pilots, it still contains specific "*energy management traps,*[6]" especially for fast jet performers[7], that must be identified and avoided during the risk assessment process.

In retrospect, I should have requested to practice both the high and low shows at the display area, allowing me to adhere to the principle of "*Train as you display, and display as you train.*"

6.2.3.4 Usage of Technology

Advanced technology, such as the Ground Collision Avoidance System (GCAS), helped me avoid a tragic accident by providing real-time and predictive alerts. This technology monitors the aircraft's altitude, speed, and trajectory, warning pilots of an imminent collision with the ground and, in some cases, taking control to initiate a recovery maneuver if the pilot fails to respond in time (Lockheed Martin, 2023).

6.2.3.5 Coordination and Communication

Clear communication between the safety observer and me was crucial for ensuring the smooth execution of the display. This coordination allows for efficient management of airspace, timely updates on weather conditions, and prompt responses to any unexpected issues.

Implementing and maintaining these risk management procedures contributed to my survival.

6.2.4 Lessons Relearned

I learned numerous valuable lessons from this incident, primarily concerning distractions, which can kill one, and risk assessment, which can save one. While these insights were fresh for me, they were not novel within the air show community regarding safety considerations.

6.2.4.1 Distraction and Interruption in Aviation

Distractions and interruptions in aviation can arise from various sources, including air traffic control (ATC), abnormal conditions, weather, or nearby traffic. Human error, often linked to distractions or preoccupations with one task while excluding another, accounts for nearly 50% of aviation accidents (Wiegmann & Shappell, 2016). In my case, a self-induced distraction led to a near-catastrophic outcome.

6.2.4.2 The Importance of Risk Assessment and Safety Margins

The incident underscores the importance of comprehensive risk assessments and adherence to safety margins for air show performers. By maintaining safe distances from the crowd and sufficient altitude to complete each maneuver, performers can reduce the likelihood of accidents and ensure the safety of both themselves and the spectators. The aviation community, particularly air show performers, must remain vigilant in their risk assessments and adhere to established safety guidelines and regulations to provide safe and entertaining aerial displays for the public.

6.2.4.3 Impact on Society

As I worked to regain control of the jet in Thessaloniki, a thought crossed my mind: How disappointed and embarrassed the Greek people would have been if I had crashed right before their eyes. Now, in hindsight, I say that it would have been devastating for my family and, indeed, for the entire Greek nation. It would have also cast a dark shadow over the Hellenic Air Force. The cause was an entirely unforeseen, self-induced distraction that would have been impossible for the accident investigation team to decipher, even if they were able to recover the jet's "black box." They might have likely attributed the accident to G-induced loss of consciousness or, as a vague factor, blamed it on pilot error.

6.3 CASE STUDY 2. ESCAPING A WOUNDED HORNET, RCAF CF-188, 2010

This remarkable survival story was chosen as a case study to illustrate the decision-making process during the critical moments leading up to an ejection or bailout from an aircraft. Fortunately, Capt. Bews confidently and promptly pulled the ejection handle (Royal Canadian Air Force, 2012), a decision that saved his life – a choice that others, in the past, failed to make, ultimately costing them their lives. It is vital to remember that the value of a person's life is immeasurable, and each individual is irreplaceable to their loved ones.

Much like tightrope walkers who rely on a safety net positioned meters above the ground to save their lives in case of a fall, air show performers flying aircraft equipped with ejection seats, bailout capabilities, or ballistic parachutes should be prepared to use these safety features to save their lives in the event of an unrecoverable situation.

6.3.1 Interview Analysis

In this section, we analyze an interview conducted with Capt. Brian Bews following the 2010 Lethbridge International Air Show CF-18 crash (*CF-18 Pilot Describes Spectacular Crash*, 2010). We will focus on the key themes emerging from his statements, such as his emotional reactions, decision-making process, and the support he received.

6.3.1.1 *Emotional Reactions*

Throughout the interview, Capt. Bews expresses various emotions in response to the crash. He conveys a sense of relief and gratitude for having survived the incident (*CF-18 Pilot Describes Spectacular Crash*, 2010). Research has shown that individuals who experience near-miss events often report feelings of relief and gratitude (Rozin & Royzman, 2001). Additionally, he exhibits humility by recognizing that even experienced pilots can encounter unexpected challenges.

6.3.1.2 *Decision-Making Process*

Capt. Bews describes his decision-making process during the emergency. He explains how his training kicked in, allowing him to make quick decisions and ultimately decide to eject from the aircraft. This aligns with research suggesting well-trained pilots can rely on automatic responses and make effective decisions under high-stress situations (Orasanu et al., 2001). For example, an expert pilot experiencing systems failures may feel they have adequate knowledge and expertise to cope with such an event and, as a result, feel low levels of stress when this occurs (Bouchard et al., 2010; Caroll et al., 2013; Schriver et al., 2008).

6.3.1.3 Resilience and Personal Growth

Despite the traumatic nature of the event, Capt. Bews demonstrates resilience in the face of adversity (*CF-18 Pilot Describes Spectacular Crash*, 2010). His ability to learn from the incident and become more cautious during flight operations illustrates the potential for personal growth following a traumatic experience (Tedeschi & Calhoun, 2004).

6.3.1.4 Support from Colleagues and the Community

In the interview, Capt. Bews expressed gratitude for the support he received from his colleagues, friends, and family (*CF-18 Pilot Describes Spectacular Crash*, 2010). Social support has been identified as a critical factor in helping individuals cope with and recover from traumatic experiences (Thoits, 2011). He acknowledges the significance of their encouragement and care in helping him process and cope with the incident's aftermath.

6.3.1.5 Reflections on the Incident

Capt. Bews shares his reflections on the incident, stating that it reminds him of the inherent risks associated with flying high-performance military aircraft. He also mentions that the incident has made him more cautious and aware of potential dangers during flight operations. This increased awareness and caution align with the concept of post-traumatic growth, where individuals learn and grow from their traumatic experiences (Tedeschi & Calhoun, 2004).

6.3.1.6 Lessons Learned

The interview highlights the importance of rigorous training, following established procedures, and the value of teamwork and support networks in overcoming high-stress situations. In addition, the insights gained from Capt. Bews' interview can inform efforts to improve pilot preparedness and resilience in emergency scenarios. Capt. Bews' experience underscores the need for continued emphasis on these aspects within the air show community to ensure pilots are well-prepared to handle emergencies and make critical decisions under pressure.

Capt. Brian Bews' interview following the 2010 Lethbridge International Air Show CF-18 crash provides valuable insights into the psychological and emotional aspects of his experience. His statements emphasize the importance of training, adherence to procedures, and support networks in successfully dealing with high-stress situations.

6.3.2 What, When, Where: The Facts

On July 23, 2010, a Canadian Forces CF-18 Hornet (see Figure 6.3) crashed during a performance at the Lethbridge International Air Show in Alberta, Canada (Royal Canadian Air Force, 2012; The Globe and Mail, 2012). The pilot, Capt. Brian Bews experienced a technical malfunction during a low-speed, low-altitude pass, causing the aircraft to lose altitude and crash. Fortunately, the pilot ejected safely before the crash, sustaining only minor injuries.

After his ejection, he landed close to the jet's fireball, but due to the strong prevailing winds and his post-ejection injury, Capt. Bews could not disconnect one of the two clips on his parachute harness (Karachalios, personal communication, 2023). Then the Sky Hawk Parachute Demonstration Team played a significant role in assisting Capt. Bews during the aftermath of the CF-18 crash (The Globe and Mail, 2012). They chased him, helping release the parachute and providing immediate first aid, highlighting the importance of teamwork, expertise, and quick thinking in emergency situations.

Capt. Bews sustained minor injuries, including compression fractures in three vertebrae. The quick actions of the pilot and the safety measures in place at the air show prevented any further injuries or damage. The investigation into the accident (Royal Canadian Air Force, 2012; The Globe and Mail, 2012) revealed a malfunction in the aircraft's engine caused by a stuck piston in the fuel control system, which led to a loss of thrust and fuel flow imbalance between the two engines, resulting in the crash.

FIGURE 6.3 Royal Canadian Air Force Demo CF-188 Hornet at RIAT 2018. Copyright: Tim Felce.

6.3.3 Why: The Psychological Factors Involved

The 2010 Lethbridge International Air Show CF-18 crash was a high-stress event for the pilot involved, Capt. Brian Bews. This case study analysis examines the psychological factors that may have influenced the pilot's decision-making and actions during the incident, mainly focusing on the ejection process. A video interview from the pilot that is available online was utilized to gather some information about his heroic survival.

6.3.3.1 Stress and Cognitive Functioning

High-stress situations, such as an aircraft malfunction, can impair cognitive functioning and decision-making processes (Driskell & Salas, 1996). High stress can lead to attentional narrowing by reducing an individual's environmental scan and thus reducing the frequency and amount of information sought by the operator (Staal, 2004). In terms of determining alternative causes of action on working memory, stress can limit situation assessment and reduce the number and quality of decision options considered (Staal, 2004; Wickens et al., 1993). Stress can increase errors and movement variability in perceptual-motor tracking tasks (Matthews & Desmond, 2002). Pilots can be faced with scenarios that can require a rapid but safety-critical information processing cycle (sensing, perceptual encoding, decision-making, action, and evaluation), and the quality of the cyclic process, as enumerated above, can be affected by stress (Entin & Serfaty, 1990; Vidulich et al., 2010).

6.3.3.2 Training, Expertise, and Automaticity

Training plays a crucial role in helping pilots manage stress and make appropriate decisions during emergencies. High training levels can lead to automaticity development, where specific actions and responses become ingrained and can be performed with minimal conscious effort (Orasanu & Martin, 1998). With consistent training coupled with real-time feedback for correction, the level of expertise increases.

Expertise can increase decisiveness and reduce decision time (Khoo & Mosier, 2005). As expertise increases, decision makers appear to be more easily able to assess which cues are relevant and need attention, as well as orient more effectively to the situation. This automaticity can enable expert pilots to identify necessary information quickly and integrate that information more effectively during high-stress situations (Wiggins et al., 2002).

6.3.3.3 Ejection Decision-Making

The decision to eject from an aircraft is complex and influenced by several psychological factors (Wright & Haynes, 2009). Pilots must quickly assess the likelihood of regaining control of the aircraft and weigh the risks of ejection against the risks of remaining in the aircraft. Factors such as self-preservation instincts, perceived control, and confidence in ejection systems can impact the decision-making process.

6.3.3.4 Capt. Brian Bews' Ejection

In the case of the Lethbridge CF-18 crash, Capt. Bews faced a rapidly deteriorating situation and decided to eject just before the aircraft impacted the ground. His successful ejection and survival may be attributed to a combination of his training, automatic responses, and quick decision-making under extreme stress.

The psychological factors affecting Capt. Brian Bews highlighted the importance of training and mental preparedness for pilots during the 2010 Lethbridge International Air Show CF-18 crash, particularly during the ejection process. By understanding these factors, aviation experts can develop more effective training programs and support systems to help air show performers cope with high-stress situations and make critical decisions to save lives.

6.3.4 How: Risk Management Strategies that Contributed to the Epic Survival

Despite the dramatic nature of the crash, it demonstrated the effectiveness of the safety measures in place, such as the ejection seat and emergency response teams, which allowed Capt. Bews survived the incident with only minor injuries.

6.3.4.1 Pilot Training and Qualifications

As Capt. Bews recalled, all the training he had received played a crucial role during those critical moments, from experiencing the loss of thrust to ejecting from the aircraft and ultimately being rescued on the ground. His extensive preparation, built on a foundation of rigorous practice and emergency training, equipped him with the skills and knowledge needed to react calmly and effectively under extreme pressure.

This experience serves as a powerful reminder of the importance of comprehensive pilot training and maintaining high qualification standards. By investing in these areas, the air show industry can work to protect both the performers and the public while fostering an enjoyable and thrilling experience for all. The combination of mastering aerial maneuvers and being prepared for emergencies ensures that pilots like Capt. Bews have the best possible chance of overcoming challenges they may face during their aerial displays.

6.3.4.2 Strict Guidelines for Air Show Performances

Capt. Bews demonstrated a commitment to safety by adhering to the minimum safe distances from the crowd, which prevented any debris from his jet from falling onto spectators. Additionally, his slow-speed pass was flown at a sufficient altitude to allow for a recovery buffer and an in-envelope ejection. These actions showcase the importance of following established guidelines to ensure the safety of both pilots and spectators during air shows.

Establishing and enforcing guidelines regarding minimum safe distances, altitude restrictions, and aircraft inspections can significantly help minimize risks to pilots and spectators alike. By adhering to these regulations, pilots can ensure that their aerial displays maintain an appropriate level of safety while still providing thrilling entertainment for the audience.

6.3.4.3 Emergency Response and Preparedness

In the aftermath of the CF-18 disaster, the Sky Hawk Parachute Demonstration Team played a crucial role in assisting Captain Bews (The Globe and Mail, 2012). They chased him swiftly, assisted with the release of his parachute, and offered rapid first assistance upon landing. This life-saving approach emphasizes the value of preparedness, competence, and quick thinking in times of crisis.

Having well-trained and equipped emergency response teams on standby is essential for minimizing the consequences of accidents when they do occur. These teams should be prepared to act swiftly and effectively in various emergency scenarios, including aircraft crashes, medical emergencies, or other unexpected events. Their presence not only provides an added layer of safety for pilots and spectators but also serves as a valuable resource for managing and mitigating potential risks during air shows.

Moreover, regular training exercises and simulations involving emergency response teams, pilots, and event organizers can help ensure seamless coordination and communication during actual emergencies. These exercises provide an opportunity to identify and address potential weaknesses in emergency response plans, ultimately enhancing the overall safety and effectiveness of the response.

The 2010 Lethbridge International Air Show CF-18 crash serves as a reminder of the inherent risks associated with high-performance military aircraft demonstrations. By identifying and addressing the contributing factors and implementing effective risk mitigation strategies, the aviation community can work together to improve the safety of air shows and prevent similar incidents in the future.

6.4 CASE STUDY 3. RESILIENCE AND SAFETY CULTURE: THE RED ARROWS' TRIUMPH OVER TRAGEDY, RAF RED ARROWS, 2010

The primary reason for choosing this accident as a case study in this book is the exemplary safety culture embodied by the Red Arrows aerobatic team of the Royal Air Force (RAF). In 2010 and 2011, following three tragic accidents that resulted in the loss of two pilots, the team demonstrated resilience, adaptability, and fortitude in recovering from such profound emotional and operational setbacks.

Moreover, the two team members involved in the accident analyzed below both remained with the Red Arrows, and no impulsive punitive actions were taken against them. The existing no-blame culture that exemplifies the team and the RAF's leadership is a model for all of us to remain committed and continue delivering exceptional performances throughout the display season.

6.4.1 What, When, Where: The Facts

On March 23, 2010 (Royal Air Force, United Kingdom, 2010), during the third sortie of the day and shortly into the second half of the Display, the Synchro Pair (see Figure 6.4), i.e., Red 6 and Red 7, collided while executing the initial level cross of

FIGURE 6.4 Red Arrows synchro pair performing a crossover at RIAT 2018. Copyright: Tim Felce.

the Opposition Barrel Roll (OBR). Red 7's fin grazed Red 6's windscreen, while the right tip of Red 7's tailplane brushed against Red 6's right wing. After the collision, Red 6's aircraft continued to descend, prompting the pilot to initiate ejection. The ejection was successful, but the pilot sustained severe injuries and was rushed to the hospital. Meanwhile, Red 7's aircraft managed to regain altitude, and the pilot diverted along with the rest of the Team to Hellenic Air Force (HAF) Heraklion Air Base, managed to perform an emergency landing despite the damage sustained during the accident.

It is important to note that the aircraft collision occurred within the boundaries of HAF Kastelli Air Base, with debris scattered across the airfield yet on the safe side of the crowd line.

6.4.2 Why: The Psychological Factors Involved

According to the investigation boards' findings (Royal Air Force, United Kingdom, 2010), several psychological factors were involved in the accident between Red 6 and Red 7. More specifically, the following are the most relevant to our readers:

6.4.2.1 Decision-Making

Both Red 6 and Red 7 had to engage in time-consuming decision-making processes due to the unfamiliar visual picture (Royal Air Force, United Kingdom, 2010, Findings para. 28.i). The lack of knowledge to determine if the presented visual cues met the escape criterion may have contributed to the accident (Royal Air Force, United Kingdom, 2010, Findings para. 29.d). In high-stress situations like aerobatic displays, decision-making under pressure can be significantly affected.

6.4.2.2 Distraction

Red 7 experienced distraction as he flew through his cross point (Royal Air Force, United Kingdom, 2010, Findings para. 29.a). This distraction may have contributed to his failure to maintain proper separation and execute the required escape maneuver.

6.4.2.3 Effect of Conditioning

Both pilots were conditioned to standard cadence and timings through repetitive training (Royal Air Force, United Kingdom, 2010, Findings para. 29.b). This conditioning may have led Red 6 to assume he had time to glance away from Red 7 before making the 'Roll go' call, affecting their decision-making processes. Moreover, the conditioning effect on both pilots (Royal Air Force, United Kingdom, 2010, Summary of Causes and Factors para. 29.b) may have influenced their decision-making processes, leading them to rely on ingrained habits instead of making adaptive decisions (Royal Air Force, United Kingdom, 2010, Summary of Causes and Factors para. 29.c).

6.4.2.4 Natural Human Reaction

Doubt, survival instinct, and natural human reactions played a crucial role in the accident (Royal Air Force, United Kingdom, 2010, Findings para. 29.e). Red 7's doubt about their separation and his instinct to avoid collision led him to bunt and fly beneath Red 6. Despite months of training, both pilots reacted instinctively rather than conditionally to the unfamiliar situation.

6.4.2.5 Biases

While the panel's investigation focused on various factors, including human factors that contributed to the accident, it is also essential to consider potential biases that may have influenced the pilots involved in the Red Arrows accident. Here are a few biases that could have affected the pilots' actions and decisions:

6.4.2.5.1 Anchoring Bias

Anchoring bias refers to the tendency to rely too heavily on an initial piece of information when making decisions (Tversky & Kahneman, 1974; C. D. Wickens et al., 2021). In this case, the pilots may have anchored their decision-making on past experiences or familiar visual cues, which could have made it difficult for them to adapt to the unique circumstances of the accident sortie.

6.4.2.5.2 Confirmation Bias

As per the investigation panel, the pilots might have also been susceptible to confirmation bias during the accident. They could have been more inclined to focus on information or visual cues supporting their preexisting beliefs or expectations about the maneuver while disregarding contradictory information.

6.4.2.5.3 Sunk Cost Fallacy and Continuation Bias

This bias occurs when individuals continue with a course of action because they have already invested resources, such as time and effort, into it (Arkes & Blumer,

1985; Salas & Maurino, 2010). In the Red Arrows accident, the pilots might have felt compelled to continue the maneuver despite recognizing potential risks, as they had already been through the second part of their display profile.

These biases, along with other human factors, could have influenced the pilots' actions and decision-making during the accident sortie, ultimately contributing to the collision between the two aircraft. Recognizing and mitigating these biases is crucial for enhancing situational awareness and decision-making in high-stakes environments like low-level aerobatic displays.

6.4.2.6 Ejection Decision

The ejection decision made by Red 6 during the accident sequence is a critical aspect to consider for all those flying in ejection seats. After the midair collision, Red 6 perceived a downward vector and sensed the windscreen shattering, which prompted the decision to initiate the ejection sequence (Royal Air Force, United Kingdom, 2010, Accident Sequence, para. 10). This decision was made just before the aircraft impacted the ground. The ejection was executed within the safe ejection envelope of the ejection seat, and all escape systems functioned correctly. Eyewitness testimony reported that the parachute opened just fully, only to slow the pilot's rapid descent toward the ground.

The ejection decision made by Red 6 demonstrates the pilot's situational awareness, quick thinking, and adaptive decision-making abilities in an emergency.

6.4.3 How: Risk Management Strategies that Contributed to the Epic Survival

The epic survival of Red 6 and Red 7 after their midair collision can be primarily attributed to effective risk management strategies. Red 6's prompt decision to eject highlights the significance of having well-trained pilots who can confidently and instinctively react and execute emergency procedures when confronted with critical situations.

6.4.3.1 Training and Supervision

The Red Arrows' success in overcoming such a challenging incident can be traced back to their comprehensive training and supervision program. This annual training program ensures that pilots are adequately prepared for the complexities and demands of their aerial displays. By participating in rigorous training exercises and receiving continuous supervision, pilots become proficient in handling a wide range of scenarios, ultimately contributing to their ability to react swiftly and effectively during emergencies.

6.4.3.2 Standard Operating Procedures

Adherence to the Team's Standard Operating Procedures is crucial in maintaining safety during displays. These procedures provide a structured framework for pilots to follow, ensuring that every maneuver is executed with precision and caution but also at a safe distance from the crowd. By consistently following the SOPs, the Red

Arrows team demonstrates a commitment to safety, which ultimately played a significant role in the successful outcome for Red 6 and Red 7 after their midair collision.

6.4.3.3 Minimum Distance from the Crowd

It is crucial to emphasize that the aircraft collision occurred within the confines of Hellenic Air Force's (HAF) Kastelli Air Base. The debris resulting from the incident was scattered throughout the airfield; however, it stayed within the safe side of the crowd line. This outcome verifies that the planning of the maneuvers was designed to prioritize crowd safety, ensuring their protection in the event of accidents like this one. This fortunate circumstance prevented any harm to people on the ground, underscoring the importance of adhering to safety regulations and ensuring proper crowd management during air shows.

Continuing safety measures and crowd control efforts, such as establishing a safe distance between the spectators and the performing aircraft, play a critical role in minimizing potential risks.

6.4.3.4 Crash Response

The aircraft collision and subsequent recovery procedures were pivotal in ensuring Red 6's safety following the accident. The prompt execution of these procedures highlights the importance of thorough training and adherence to safety protocols.

The crash response team was activated immediately after the collision. Their swift action and readiness to address emergencies helped to minimize potential hazards and facilitate the safe recovery of Red 6.

Prompt medical support was provided to Red 6 upon landing, ensuring that any potential injuries were promptly addressed and treated. The availability of medical personnel and their ability to respond quickly in such situations is critical for pilots' well-being in accidents.

The combination of these crash and recovery procedures, along with the pilots' adherence to safety protocols and exceptional skills, played a significant role in caring for Red 6 after the accident.

6.4.3.5 Safety Culture

The Red Arrows aerobatic team of the Royal Air Force (RAF) exemplifies an outstanding safety culture, making it an ideal case study for analyzing accident recovery and resilience. In 2010 and 2011, the team faced three tragic accidents: the first occurred on March 23, 2010, as presented in the previous pages; the second accident happened on August 20, 2011, when Flight Lieutenant Jon Egging's Hawk T1 aircraft crashed after a display at the Bournemouth Air Festival, resulting in his death (BBC News, 2011); the third incident took place on November 8, 2011, when Flight Lieutenant Sean Cunningham was fatally injured after being ejected from his aircraft while on the ground (Drury, 2011). Despite the immense emotional and operational challenges, the Red Arrows displayed remarkable resilience, adaptability, and fortitude in overcoming these setbacks.

A critical aspect of the Red Arrows' safety culture is their emphasis on a no-blame environment. Following the accident discussed in the case study, both team members remained with the team, and there were no impulsive punitive actions taken against them. This approach underscores the team's commitment to learning from incidents, focusing on improvement and growth rather than assigning blame.

The leadership of the RAF and the Red Arrows team recognizes the importance of fostering a culture that encourages open communication, accountability, and continuous learning. By prioritizing safety and promoting a supportive atmosphere, they create an environment where team members can address concerns, share experiences, and learn from one another. This mindset contributes to the overall safety culture and enables the team to maintain exceptional performance throughout the display season.

The Red Arrows' safety culture serves as a model for organizations across various industries. Key elements that other teams can adopt include comprehensive training and supervision, open communication, continuous learning and improvement, a blame culture, and leadership commitment. Organizations can promote a resilient and adaptable environment that supports continuous improvement and exceptional performance by adopting and integrating these principles into their safety culture.

6.5 CASE STUDY 4. AN UNFORESEEN ENCOUNTER: EPIC SURVIVAL AMONG AIR SHOW PERFORMERS IN VINTAGE WARBIRDS, 2008

The current case study of an epic survival showcases the intricacies related to air show performers flying warbirds, mainly focusing on a 2008 incident involving a Vickers Spitfire MK9 and a Hurricane aircraft, see Figure 6.5 (National Transport Safety Board, 2008). The accident occurred during an air show at Scholes International Airport in Galveston, Texas, where the two aircraft collided while taxiing postlanding, causing substantial damage but miraculously resulting in no fatalities or injuries.

6.5.1 What, When, Where: The Facts

On April 26, 2008, a Vickers Spitfire MK9, registered as N308WK, participated in an aerial demonstration alongside a Hurricane aircraft and several other planes at Scholes International Airport, Texas, USA. After completing the performance, both the Spitfire and Hurricane were preparing to land when a miscommunication between the pilots led to the accident.

The lead pilot had radioed that everyone would land and exit the runway at its end, which the Spitfire pilot took as a cue to apply only light braking after landing. However, the Hurricane aircraft drifted to the left side of the runway and ground looped, attempting to return to the right side when it was struck by the Spitfire. The pilot of the Hurricane reported that the right brake had failed to respond, which

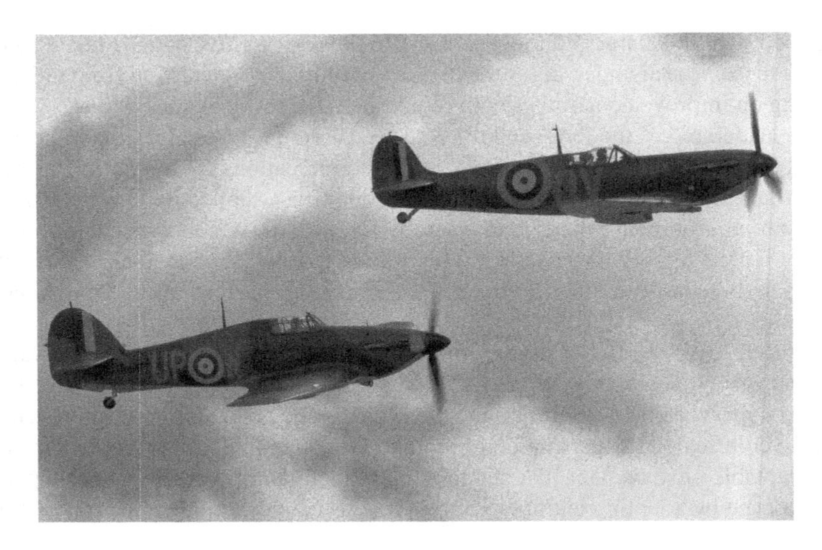

FIGURE 6.5 Spitfire and Hurricane in formation. Copyright: Tim Felce.

contributed to the collision. In contrast, the Spitfire pilot reported that during a three-point landing in the Spitfire, forward visibility was restricted.

6.5.2 Why: The Psychological Factors Involved

Psychological factors played a significant role in the incident involving the Vickers Spitfire MK9 and the Hurricane aircraft during the air show. In particular, factors such as stress, workload, communication errors, maintaining vigilance, and expectation bias contributed to the accident.

6.5.2.1 Stress

Air show performances often place pilots under significant stress, as they are required to execute complex maneuvers while ensuring the safety of themselves and others. The high-stress environment at a critical phase of flight, i.e., landing, may have affected the pilots' decision-making and response times, leading to the collision.

6.5.2.2 Workload and Task Saturation

Pilots participating in air shows often face high workloads because they need to maintain situational awareness, manage their aircraft, and coordinate with other performers. The high workload during air show events, including maintaining formation and executing complex maneuvers, may have contributed to task saturation for the pilots (Wickens, 2002). This can result in reduced attention to critical tasks, such as maintaining directional control during roll out for the Hurricane pilot or monitoring the position of other aircraft for the Spitfire pilot.

6.5.2.3 Communication Errors

Effective communication is crucial in aviation, particularly during high-stakes situations like air shows. In this incident, the communication between the lead pilot and the Spitfire and Hurricane pilots appears to have been unclear, contributing to the misunderstanding about the intended landing and runway exit procedures. Miscommunication can lead to errors and accidents, emphasizing the need for clear and concise information exchange among pilots.

6.5.2.4 Maintaining Vigilance

Maintaining vigilance refers to the act of keeping a continuous state of alertness, watchfulness, and attentiveness to one's surroundings, potential hazards, or changes in a given situation. In the context of safety-critical tasks, such as aviation and air show performances, maintaining vigilance is crucial for ensuring safety and effective performance(Thomson et al., 2015). In the context of the Spitfire and Hurricane aircraft incident, the challenges of maintaining vigilance could have played a role in the accident by contributing to lapses in attention, overconfidence, and a reduced sense of urgency.

6.5.2.4.1 Lapses in Attention

Air show performers face unique challenges in maintaining vigilance during their performances, which is crucial for ensuring safety and delivering a successful show. Pilots need to be continuously aware of their environment, aircraft systems, and the precise execution of their maneuvers. As Cheyne (2010) highlights, attention lapses can occur in situations of both high and low workloads, with humans naturally struggling to sustain attention on a single task.

In the context of air show performers, low workload scenarios can be particularly concerning, as boredom may cause pilots to lose focus on the critical aspects of their performance (Thomson et al., 2015). This waning vigilance could lead to decreased situational awareness and errors in executing maneuvers, potentially putting the pilot and spectators at risk.

In this case, both pilots may have been less vigilant about their ability to execute the landing and taxi procedures, leading them to pay less attention to the actions and intentions of the other aircraft.

6.5.2.4.2 Overconfidence

Lack of maintaining vigilance can result in overconfidence, where pilots believe that their skills and experience are sufficient to handle any situation, even in the face of potential hazards (Wiegmann & Shapell, 2016). The Spitfire pilot may have been overconfident in his ability to judge the distance between his aircraft and the Hurricane, leading him to apply only light braking after landing. Similarly, the Hurricane pilot may have been overconfident in the reliability of the aircraft's braking system, not anticipating the reported brake failure.

6.5.2.4.3 Reduced Sense of Urgency

Failure to sustain vigilance can also create a reduced sense of urgency, causing pilots to underestimate the need for timely and appropriate actions (Salas & Maurino, 2010). In this incident, both pilots may have been complacent about the need to exit the runway quickly and safely, leading to a slower response when the Hurricane began drifting and ground looping and not communicating the situation to the trailing aircraft. This lack of urgency could have contributed to Spitfire's collision with the Hurricane.

In summary, failure to maintain vigilance may have influenced the Spitfire and Hurricane accident by causing lapses in attention, overconfidence, and a reduced sense of urgency among the pilots. To mitigate the risks associated with a lack of maintaining vigilance, it is essential to implement strategies that support sustained attention, such as varied stimuli, effective workload management, and appropriate levels of automation. Additionally, fostering a culture of safety that emphasizes the importance of vigilance, thorough planning, and adherence to standard operating procedures can help to maintain attention and enhance overall performance in safety-critical tasks, ultimately reducing the risk of similar accidents in the future.

6.5.2.5 Expectation Bias

The Spitfire pilot expected the Hurricane to exit the runway at the end, which may have led to reduced situational awareness and a slower reaction time when the Hurricane failed to do so (Wickens et al., 2021). Therefore, it is essential for pilots, as well as for air bosses, to continuously assess and reassess the situation during critical phases of flight, such as landing and taxiing, and to avoid relying solely on expectations while communicating any unexpected situations.

6.5.3 How: Risk Management Strategies that Contributed to the Epic Survival

While the incident involving the Spitfire and Hurricane aircraft resulted in epic survival, specific risk management strategies may have played a role in preventing fatalities and injuries. These strategies are essential components of safety management in aviation, aimed at identifying, assessing, and mitigating potential hazards (Wiegmann & Shapell, 2016). The following risk management strategies may have contributed to the positive outcome of the incident:

6.5.3.1 Effective Crew Resource Management

Crew resource management (CRM) involves the efficient use of all available resources to ensure safe and successful flight operations (Kanki et al., 2019). While the incident involved miscommunication between pilots, other aspects of CRM, such as situational awareness, decision-making, and problem-solving, may have helped minimize the consequences of the collision, especially during the postcrash response.

6.5.3.2 Emergency Response Preparedness

Air shows and aviation events typically involve comprehensive emergency response planning, ensuring that medical and rescue personnel are available to address potential incidents (Coppola, 2020). The rapid response of emergency crews following the Spitfire and Hurricane collision may have contributed to the absence of injuries and fatalities.

6.6 CASE STUDY 5. A NEAR DISASTER AT THE PARIS AIR SHOW: THE 1989 MIG-29 BIRD STRIKE INCIDENT

The 1989 incident at the Paris Air Show in Le Bourget involved a Mikoyan MIG-29 (see Figure 6.6), highlighting the persistent threat that bird strikes pose to air show performers over the years.

6.6.1 What, When, Where: The Facts

On June 8, 1989, at the international air show in Le Bourget, the latest Soviet fighter MIG-29 crashed during a demonstration flight. During the flight, a bird was ingested into the aircraft's right engine, causing it to lose power. The pilot, Anatoly Kvochur, attempted to continue flying using the remaining left engine, but the asymmetrical thrust made it difficult to maintain control of the aircraft. As the aircraft rapidly

FIGURE 6.6 MIG-29 performing at RIAT 2016. Copyright: Tim Felce.

descended toward the ground, Kvochur ejected just seconds before the crash and sustained only minor injuries. The bird strike was identified as the primary cause of the engine failure and subsequent crash.

In preparing this book, an online interview with Anatoly Kvochur, the pilot of the MIG-29 involved in the 1989 Paris Air Show crash in Le Bourget, was accessed and translated from Russian into English. Interestingly, Kvochur noted that the two-second span between the ejection sequence activation and his landing on the ground felt very different for him compared to the observers. Kvochur experienced a sense of time compression, where more events and actions seemed to transpire within the exact brief moment.

Upon examining the video footage, one can see flames erupting from the right engine's nozzle. The aircraft nosedives rapidly, making it evident that the jet is beyond recovery. At an altitude of 92 meters, the ejection process occurs, and the pilot can be seen falling at the same speed as the plane, approximately 40 meters per second. The parachute canopy has insufficient time to fully deploy and protect Kvochur. Nevertheless, the plane's explosion produces a shockwave that inflates the parachute canopy, reducing the pilot's descent speed to 12 meters per second. This allows for a gentler landing on the ground, enabling Kvochur to survive the incident.

6.6.1.1 Bird Strikes at Air Shows

Bird strikes pose significant risks to aviation safety, as they can lead to engine damage, loss of control, and even crashes (Juračka et al., 2021; Metz et al., 2020). The risks are amplified for air show performers, often flying at low altitudes and high speeds, leaving little room for error or recovery.

6.6.1.1.1 Engine Damage

Bird strikes can cause severe damage to aircraft engines, especially when birds are ingested into the engine's intake (Allan & Orosz, 2001). Depending on the bird's size and the aircraft's speed, bird strikes can lead to a loss of engine power, compressor stalls, or even engine failure (Dolbeer & Eschenfelder, 2003).

6.6.1.1.2 Structural Damage

In addition to engine damage, bird strikes can cause structural damage to the aircraft, such as damage to the wings, tail, windshield, or fuselage (Thorpe, 2003). Such damage can compromise the aircraft's structural integrity and flight control, posing a significant safety risk.

6.6.1.1.3 Air Show Performers

Air show performers are at increased risk of bird strikes due to the low altitudes, high speeds, and complex aerobatic maneuvers involved in their performances. These factors reduce the margin for error and increase the potential for accidents in the event of a bird strike.

6.6.1.1.4 Examples and Statistics

According to the Federal Aviation Administration (FAA), there were over 15,000 reported bird strikes in the United States in 2021 (Federal Aviation Administration,

2023). Moreover, in the 2022 display season, three out of the nine recorded air show accidents were related to bird strikes, with one leading to the safe ejection of a Turkish Stars pilot (Daily Sabah, 2022).

6.6.2 Why: The Psychological Factors Involved

Several psychological factors could be related to the MIG-29 crash at the Paris Air Show, with the main to include time compression, tunnel vision, and cognitive overload.

6.6.2.1 Time Compression

The psychological factor Kvochur may have experienced, known as time compression, is a phenomenon where an individual perceives time as moving more slowly during intense or high-stress situations (Arstila, 2012). This altered perception allows the person to feel that more events and actions are occurring within the same short time frame. In the context of Kvochur's experience, the stress and adrenaline rush during the ejection sequence and rapid descent likely triggered this time compression sensation.

This phenomenon is considered an adaptive response to stress, enabling individuals to process information more rapidly and make quicker decisions when faced with life-threatening situations (Stetson et al., 2007). The brain's heightened state of arousal during such events enhances sensory processing and cognitive function, making it seem as if time is slowing down, allowing the individual to react more effectively to the situation (Wittmann & Van Wassenhove, 2009). This altered perception of time can be critical in high-pressure scenarios, such as the one Kvochur faced, as it may contribute to a person's ability to survive and navigate through dangerous circumstances.

6.6.2.2 Tunnel Vision

Under extreme stress, individuals may experience tunnel vision, where their focus narrows on a specific aspect of the situation, causing them to miss crucial peripheral information (Eysenck et al., 2007).

6.6.2.3 Cognitive Overload

High-stress situations can lead to cognitive overload, where an individual struggles to process and manage the large amount of information coming in. This can impair decision-making and performance (Sweller, 1994).

6.6.3 How: Risk Management Strategies that Contributed to the Epic Survival

Several risk management strategies played a crucial role in the epic survival of pilot Anatoly Kvochur during the MIG-29 crash at the Paris Air Show. These strategies included the advanced ejection system of the jet, Kvochur's training and experience, the emergency response, and his adherence to air show safety regulations.

6.6.3.1 Advanced Ejection System

The MIG-29's ejection system, designed to remove the pilot quickly and effectively from the aircraft in emergency situations during extreme flying conditions, played a critical role in Kvochur's survival. The system's efficiency and reliability ensured that Kvochur was ejected safely and away from the aircraft, minimizing the risk of injury during the crash.

6.6.3.2 Pilot Training and Experience

Kvochur's extensive training and experience as a test pilot likely contributed to his ability to remain calm and make critical decisions during high-pressure situations. His expertise and skills allowed him to react promptly and correctly to the emergency, ultimately contributing to his survival.

6.6.3.3 Emergency Response Team

The swift response of the emergency crews on the ground at the air show ensured that Kvochur received immediate medical attention, further increasing his chances of survival. Their presence and preparedness were instrumental in providing prompt and effective care, minimizing the potential for further complications.

6.6.3.4 Safety Regulations and Procedures

The air show's safety regulations and procedures helped to ensure that Kvochur and the spectators were kept as safe as possible during the event. These guidelines likely included provisions for aircraft separation, flight demonstrations, and emergency response protocols. By adhering to these safety measures, the risk of injury or harm to both the pilot and spectators was significantly reduced.

6.6.3.5 Wildlife Management in Air Shows

Based on the high frequency of bird strike occurrences during air shows (Barker, 2020a), including the 1989 MIG-29 crash, implementing effective wildlife management strategies in air show settings is crucial. Wildlife management procedures related to FDDs and air show organizers can be found in detail on the U.S. Air Force Safety Center's website (U.S. Air Force, 2023). Below we will focus on suggestions related to air show performers:

6.6.3.5.1 Timing and Scheduling

Be aware if the air show has been scheduled during times when bird activity is high, such as migratory seasons or peak feeding times. Therefore, avoid scheduling air shows during early mornings and late afternoons when bird activity is typically high to minimize the risk of bird strikes.

6.6.3.5.2 Bird Aircraft Strike Hazard (BASH)

Incorporate bird aircraft strike hazard (BASH)) into your personal risk assessment. In addition, don't hesitate to request the FDD to include the latest Bird Watch

Condition code in their safety briefings to emphasize potential risks associated with bird strikes during the air show (U.S. Air Force, 2018).

By implementing these wildlife management strategies, flight directors and air show organizers can help minimize the risk of bird strikes and contribute to a safer environment for performers and spectators alike.

6.7 CASE STUDY 6. EXCEPTIONAL TEAMWORK: WING WALKER DITCHING AT UK, 2021

The final and most captivating case study of epic survivals showcases the remarkable teamwork between air show performers, including a wing walker and pilot. The EAC recognized their exceptional response to an emergency, highlighting their exemplary coordination and performance under such a demanding situation.

6.7.1 WHAT, WHEN, WHERE: THE FACTS

On September 4, 2021, a Boeing A75N1 Stearman operated by AeroSuperBatics Wingwalkers (see Figure 6.7) experienced an accident during an aerobatic wing-walking display at the Bournemouth Air Festival in the United Kingdom (Air Accidents Investigation Branch, 2022). While performing the display over the sea,

FIGURE 6.7 AeroSuperBatics wing walker performing over the wing. Copyright: Tim Felce.

the aircraft encountered reduced engine power. In response, the pilot halted the display routine and flew along the coast, intending to return to Bournemouth Airport. The wing walker moved back to her seat in the front cockpit.

Though the aircraft initially maintained its altitude, the engine ultimately lost all power, prompting the pilot to ditch the aircraft at the entrance to Poole Harbor. Upon contacting the water, the aircraft flipped over. However, both occupants managed to exit the aircraft unaided. The aircraft was written off, as it sustained damage beyond repair. There were no fatalities or other casualties reported in this incident.

6.7.2 Why: The Psychological Factors Involved

In the case of the Stearman accident, both the wing walker and pilot likely experienced several psychological factors that influenced their actions and reactions during the emergency. These factors include stress and arousal, situation awareness, decision-making under pressure, effective teamwork, and potential cognitive biases.

6.7.2.1 Stress and Arousal

During the time from the engine failure to the crash landing, both the pilot and wing walker would have experienced heightened stress and arousal, which can impact cognitive processes, decision-making, and performance (Driskell & Salas, 1996). This heightened state can lead to tunnel vision, in which an individual focuses solely on a specific aspect of a situation, potentially ignoring other critical information.

6.7.2.2 Situation Awareness

Both the pilot and wing walker needed to maintain a high level of situation awareness (SA) during their performance. SA involves understanding the current situation, projecting future states, and making informed decisions based on this understanding (Endsley, 1995). Despite the challenges they faced, the pilot effectively maintained an elevated level of SA. He was aware of the nearest airfield to land, maintained control of the aircraft, analyzed the situation, and took all the necessary actions to ensure a safe return for himself and the wing walker.

6.7.2.3 Decision-Making Under Pressure

In high-pressure situations, such as the one experienced during the accident, individuals often have to make rapid decisions with limited information. This can lead to decision-making biases and errors, such as the availability heuristic, where people make judgments based on the ease with which relevant examples come to mind (Kahneman & Tversky, 1979). However, the pilot's decision-making process was exemplary from the moment he initially faced the engine malfunction until landing, ensuring the safety of both himself and the wing walker.

6.7.2.4 Effective Teamwork

The successful coordination between the pilot and wing walker highlights the importance of effective teamwork during an emergency. This includes clear communication, cooperation, and trust among team members (Salas et al., 2008). Their effective teamwork ensured the avoidance of coordination errors and adverse outcomes, ultimately contributing to their safe return.

6.7.2.5 Cognitive Biases

In high-stress situations, individuals may be more prone to cognitive biases, such as confirmation bias, seeking information that confirms preexisting beliefs, and anchoring bias, relying too heavily on an initial piece of information (Tversky & Kahneman, 1974). These biases can negatively impact decision-making and situational awareness during an emergency. However, the pilot and wing walker managed to overcome these potential biases, maintaining a clear focus on the task at hand and making sound decisions that ensured their safety.

6.7.3 How: Risk Management Strategies that Contributed to the Epic Survival

The investigation by the Air Accidents Investigation Branch (AAIB) determined that the reduction in engine power occurred due to the failure of the oil inlet pipe, which resulted in the loss of the oil supply to the engine (Air Accidents Investigation Branch, 2022). Nevertheless, several risk management strategies the wing walker and the pilot applied contributed to this epic survival, including training and experience, effective communication protocols, contingency planning, and proper safety equipment.

6.7.3.1 Training and Experience

The pilot's quick response and ability to evaluate the situation and make decisions in a rapidly changing and potentially life-threatening situation demonstrate the importance of training and experience. The wing walker and pilot were well-rehearsed in handling emergencies, which contributed to the successful outcome of the accident. Moreover, the pilot was familiar with the emergency procedures and thoroughly understood the aircraft they were flying.

6.7.3.2 Communication Protocols and Teamwork

Communication between the wing walker and pilot was essential in responding effectively to the emergency. Proper communication protocols between the wing walker and pilot were clearly established and rehearsed before the air show performance. They ensured their ability to work as a team and respond to such a critical emergency. The ability to communicate effectively, cooperate, and trust each other likely played a significant role in their survival (Salas et al., 2008).

6.7.3.3 Contingency Planning

Before the air show, the pilot and wing walker may have discussed and practiced emergency procedures, including what to do in the event of an engine failure or other critical situations. This contingency planning likely contributed to their ability to react effectively during the accident.

6.7.3.4 Safety Equipment

Proper equipment, such as safety harnesses and helmets, reduced the risk of injury in the accident, especially when the airplane flipped over after the ditching. The use of appropriate safety gear ensured that both the pilot and wing walker were protected from potential injuries during the emergency, further contributing to their epic survival.

In summary, the combination of training and experience, effective communication protocols, contingency planning, and proper safety equipment contributed to the epic survival of the pilot and wing walker during the accident. These risk management strategies demonstrate the importance of preparation, teamwork, and resilience in the face of unexpected challenges during high-stakes events like air shows.

NOTES

1. A "low show" in the context of a flying display or air show refers to a modified performance that is carried out when the weather conditions are not ideal for the standard, full display. In the given scenario, the low show has specific minimum weather requirements to ensure the safety of the pilots and spectators. These requirements include:

   Minimum cloud ceiling: 3,500 feet. This means that the lowest altitude at which clouds form (the cloud base) must be at or above 3,500 feet. This allows the pilots to perform their low show routine without the risk of entering clouds, which would severely limit their visibility and ability to maintain visual contact with the ground and other aircraft.

   Visibility: More than 5 kilometers. Visibility refers to the distance at which objects can be clearly seen. In this case, the minimum visibility required for the low show is more than 5 kilometers, ensuring that pilots can maintain a safe distance from the audience while performing their maneuvers.

   By setting these specific minimum weather requirements for the low show, the event organizers can ensure that display pilots can safely perform their modified routine without compromising safety, even in less than ideal weather conditions.
2. A foul line in airshows is a designated safety boundary that separates the performance area from the spectator area. This line is established to maintain a safe distance between the aircraft and the audience, ensuring that any potential risks associated with the aerial display do not harm spectators.

   The exact distance of the foul line from the audience can vary depending on the type and size of the aircraft performing, as well as the nature of the maneuvers being executed. Generally, the faster and more powerful the aircraft, the further away the foul line will be from the audience.

   Pilots performing in airshows are required to remain on the performance side of the foul line at all times during their display, and they must not cross it. This rule helps to minimize the risk of accidents and keeps both the pilots and the audience safe during the event. In addition, airshow organizers often have emergency response teams and safety measures in place to handle any unforeseen incidents that may occur during the performance.
3. Peripheral vision is more sensitive to motion and provides a broader, less detailed view of the surroundings, while focal vision is responsible for detecting and processing detailed visual information (Wickens et. al, 2021).

4. A display line in air shows is a virtual boundary established by the national aviation authorities to maintain a safe distance between air show performers and the audience. This imaginary straight line separates the performance area from the spectator area, and its distance from the crowd varies depending on the type of aircraft and the nature of the maneuvers being performed.
5. The minimum altitude, at which a pilot is allowed to perform specific maneuvers, provides a safety buffer between the aircraft and the ground, reducing the risk of accidents due to pilot error or unexpected factors such as aircraft malfunctions or changes in weather conditions. The hard deck for the HAF F-16 demonstration team may vary depending on the specific maneuver being performed, but generally, it is set at a minimum of 100 feet above ground level (AGL) for high speed passes and 500 feet AGL for aerobatic maneuvers.

   The display line refers to the minimum distance pilots must maintain from spectators during their performance. This helps to ensure the safety of both the crowd and the pilot by providing a buffer zone in case of unexpected events or errors. For the HAF F-16 demonstration team, the display line is typically set at 500 to 1,500 feet from the crowd, depending on the type of maneuver being performed and the specific airshow regulations.
6. Energy management traps refer to situations or maneuvers during a flight that can lead to an undesirable or unsafe energy state for the aircraft. This can involve losing too much airspeed or altitude or entering a high angle of attack that could result in an aerodynamic stall or other hazardous conditions.
7. Energy management traps for fast jet performers are closely related to altitude and airspeed. The excessive thrust available could lead to high speeds, which in turn results in larger turning radiuses, requiring more space to recover from the ground. Therefore, proper energy management planning involves timely adjustments of power settings to maintain the energy levels inside the safe windows of energy (airspeed and altitude), the so called "energy gates." For example, during a loop, a pilot may need to reduce the throttle, i.e., deselect the afterburner, to avoid excessive airspeed, which would otherwise increase the turning radius and make recovery more difficult.

LIST OF REFERENCES

Air Accidents Investigation Branch. (2022). *AAIB investigation to Boeing A75N1(PT17) Stearman, N707TJ* (Bulletin - Field Investigation No. 27642; p. 19). UK Air Accidents Investigation Branch. https://www.gov.uk/aaib-reports/aaib-investigation-to-boeing-a75n1-pt17-stearman-n707tj

Allan, J., & Orosz, A. (2001). The costs of bird strikes to commercial aviation. *Proceedings of Bird Strike, 2001.*

Anderson, J. R. (1983). *Acquisition of proof skills in geometry* (R. S. Michalski, J. G. Carbonell, & T. M. Mitchell, Eds., pp. 191–219). Springer. https://doi.org/10.1007/978-3-662-12405-5_7

Karachalios M. (2023). *Interview with a witness at the demonstration CF-18 Hornet accident of July 23, 2010* (M. Karachalios, Interviewer) [In-person Interview].

Arkes, H. R., & Blumer, C. (1985). The psychology of sunk cost. *Organizational Behavior and Human Decision Processes, 35*, 124–140. https://doi.org/10.1016/0749-5978(85)90049-4

Arstila, V. (2012). Time slows down during accidents. *Frontiers in Psychology, 3*. https://doi.org/10.3389/fpsyg.2012.00196

Barker, D. (2020a). *Anatomy of airshow accidents* (1st ed., Vol. 1). International Council of Air Shows, International Air Show Safety Team.

Barker, D. (2020b, December). *Air show accidents: A statistical and preventative perspective.* ICAS Convention. https://www.youtube.com/watch?v=29zT0edEAUQ&t=877s

BBC News. (2011, August 20). Red Arrows pilot dies in Bournemouth Air Festival crash. *BBC News.* https://www.bbc.com/news/uk-england-14602900

Bock, O., Wechsler, K., Koch, I., & Schubert, T. (2021). Dual-task interference and response strategies in simulated car driving: Impact of first-task characteristics on the psychological refractory period effect. *Psychological Research, 85*(2), 568–576. https://doi.org/10.1007/s00426-019-01272-5

Bouchard, S., Baus, O., Bernier, F., & McCreary, D. R. (2010). Selection of key stressors to develop virtual environments for practicing stress management skills with military personnel prior to deployment. *Cyberpsychology, Behavior and Social Networking, 13*(1), 83–94. https://doi.org/10.1089/cyber.2009.0336

Caroll, M., Padron, C. K., DeVore, L., Winslow, B., Hannigan, F. P., Murphy, J. S., & Squire, P. (2013, December 1). *Simulation training approach for small unit decision-making under stress.* Interservice/Industry Training, Simulation, and Education Conference (I/ITSEC), Orlando, FL.

CF-18 pilot describes spectacular crash. (2010, August 17). YouTube. https://www.youtube.com/watch?v=anz4bBsZ2WM

Cheyne, A. (2010). Attention Lapses. In *The Corsini encyclopedia of psychology* (pp. 1–1). John Wiley & Sons, Ltd. https://doi.org/10.1002/9780470479216.corpsy0095

Chira, I., Adams, M., & Thornton, B. (2008). *Behavioral bias within the decision making process* (SSRN Scholarly Paper No. 2629036). https://papers.ssrn.com/abstract=2629036

CNN. (2010, March 26). *Greece's financial crisis explained.* http://www.cnn.com/2010/BUSINESS/02/10/greek.debt.qanda/index.html

Coppola, D. P. (2020). *Introduction to international disaster management* (4th ed.). Elsevier. https://www.elsevier.com/books/introduction-to-international-disaster-management/p-coppola/978-0-12-817368-8

Daily Sabah. (2022, December 6). *Military plane crashes in central Türkiye, pilot survives.* Daily Sabah. https://www.dailysabah.com/turkey/military-plane-crashes-in-central-turkiye-pilot-survives/news

Dolbeer, R. A., & Eschenfelder, P. (2003). Amplified bird-strike risks related to population increases of large birds in North America (pp. 49–67).

Driskell, J. E., & Salas, E. (Eds.). (1996). *Stress and human performance* (pp. xiv, 314). Lawrence Erlbaum Associates, Inc.

Drury, I. (2011, November 9). Second tragedy for Red Arrows after pilot is killed in freak jet "ejector seat" accident at air base. *Mail Online.* https://www.dailymail.co.uk/news/article-2059004/Red-Arrows-accident-Pilot-Sean-Cunningham-killed-RAF-Scampton.html

Endsley, M. R. (1995). Toward a theory of situation awareness in dynamic systems. *Human Factors, 37*(1), 32–64. https://doi.org/10.1518/001872095779049543

Entin, E. E., & Serfaty, D. (1990). *Information gathering and decision making under stress* (Technical Report No. DA218233; AD-E501191; TR-454). https://apps.dtic.mil/sti/citations/ADA218233

Eysenck, M. W., Derakshan, N., Santos, R., & Calvo, M. G. (2007). Anxiety and cognitive performance: Attentional control theory. *Emotion (Washington, D.C.), 7*(2), 336–353. https://doi.org/10.1037/1528-3542.7.2.336

Fagot, C., & Pashler, H. (1992). Making two responses to a single object: Implications for the central attentional bottleneck. *Journal of Experimental Psychology. Human Perception and Performance, 18*(4), 1058–1079. https://doi.org/10.1037//0096-1523.18.4.1058

Federal Aviation Administration. (2023). *FAA wildlife strike database*. FAA Wildlife Strike Database. https://wildlife.faa.gov/home

Fu, W.-T., & Gray, W. D. (2006). Suboptimal tradeoffs in information seeking. *Cognitive Psychology, 52*, 195–242. https://doi.org/10.1016/j.cogpsych.2005.08.002

Harris, D. (2011). *Human performance in the flight deck*. CRC Press.

Held, M., Rieger, J. W., & Borst, J. P. (2022). Multitasking while driving: Central bottleneck or problem state interference? *Human Factors*, 187208221143857. https://doi.org/10.1177/00187208221143857

Jung, K., Ruthruff, E., Tybur, J. M., Gaspelin, N., & Miller, G. (2012). Perception of facial attractiveness requires some attentional resources: Implications for the "automaticity" of psychological adaptations. *Evolution and Human Behavior, 33*(3), 241–250. https://doi.org/10.1016/j.evolhumbehav.2011.10.001

Juračka, J., Chlebek, J., & Hodaň, V. (2021). Bird strike as a threat to aviation safety. *Transportation Research Procedia, 59*, 281–291. https://doi.org/10.1016/j.trpro.2021.11.120

Kahneman, D., & Tversky, A. (1979). Prospect theory: An analysis of decision under risk. *Econometrica, 47*(2), 263–291. https://doi.org/10.2307/1914185

Kanki, B., Anca, J., & Chidester, T. (2019). *Crew resource management* (3rd ed.). Academic Press. https://www.elsevier.com/books/crew-resource-management/kanki/978-0-12-812995-1

Khoo, Y.-L., & Mosier, K. (2005). Searching for cues: An analysis on factors effecting the decision making process of regional airline pilots. *Proceedings of the Human Factors and Ergonomics Society Annual Meeting, 49*(3), 578–581. https://doi.org/10.1177/154193120504900377

Lien, M.-C., Ruthruff, E., & Johnston, J. C. (2006). Attentional limitations in doing two tasks at once: The search for exceptions. *Current Directions in Psychological Science, 15*, 89–93. https://doi.org/10.1111/j.0963-7214.2006.00413.x

Lockheed Martin. (2023). *Automatic ground collision avoidance system (Auto GCAS)*. Lockheed Martin. https://www.lockheedmartin.com/en-us/products/autogcas.html

Loukopoulos, L. L., Dismukes, R. K., & Barshi, I. (2010). The multitasking myth: Handling complexity in real-world operations. *Applied Cognitive Psychology, 24*(7), 1046–1047. https://doi.org/10.1002/acp.1675

Matthews, G., & Desmond, P. A. (2002). Task-induced fatigue states and simulated driving performance. *The Quarterly Journal of Experimental Psychology. A, Human Experimental Psychology, 55*(2), 659–686. https://doi.org/10.1080/02724980143000505

Metz, I. C., Ellerbroek, J., Mühlhausen, T., Kügler, D., & Hoekstra, J. M. (2020). The bird strike challenge. *Aerospace, 7*(3), Article 3. https://doi.org/10.3390/aerospace7030026

National Transport Safety Board. (2008). *Aviation investigation final report: Hurricane and Spitfire 26 Apr 2008* (Aircraft Accident Investigation Report DFW08LA118; p. 12). National Transport Safety Board. https://aviation-safety.net/wikibase/17697

Orasanu, J., & Martin, L. (1998). Errors in aviation decision making: A factor in accidents and incidents. *HESSD*. https://www.dcs.gla.ac.uk/~johnson/papers/seattle_hessd/judithlynne-p.pdf

Orasanu, J., Martin, L., & Davidson, J. (2001). Cognitive and contextual factors in aviation accidents: Decision errors. In *Linking expertise and naturalistic decision making* (pp. 209–225). Lawrence Erlbaum Associates Publishers. https://doi.org/10.4324/9781410604200

Royal Air Force, United Kingdom. (2010). *Service inquiry report into the Red Arrows accident in Crete on 23 March 2010* (p. 41) [Corporate report]. Royal Air Force, United Kingdom. https://www.gov.uk/government/publications/service-inquiry-report-into-the-red-arrows-accident-in-crete-on-23-march-2010

Royal Canadian Air Force. (2012, December 12). *Flight safety investigation report: Demonstration CF-18 accident of July 23, 2010*. Flight Safety Investigation Report: Demonstration CF-18 Accident of July 23, 2010. https://www.canada.ca/en/department-national-defence/maple-leaf/rcaf/migration/2013/flight-safety-investigation-report-demonstration-cf-18-accident-of-july-23-2010.html

Rozin, P., & Royzman, E. B. (2001). Negativity bias, negativity dominance, and contagion. *Personality and Social Psychology Review*, *5*(4), 296–320. https://doi.org/10.1207/S15327957PSPR0504_2

Salas, E., Cooke, N. J., & Rosen, M. A. (2008). On teams, teamwork, and team performance: Discoveries and developments. *Human Factors*, *50*(3), 540–547. https://doi.org/10.1518/001872008X288457

Salas, E., & Maurino, D. (Eds.). (2010). *Human factors in aviation* (2nd ed.). Academic Press. https://www.elsevier.com/books/T/A/9780123745187

Schriver, A. T., Morrow, D. G., Wickens, C. D., & Talleur, D. A. (2008). Expertise differences in attentional strategies related to pilot decision making. *Human Factors*, *50*(6), 864–878. https://doi.org/10.1518/001872008X374974

Staal, M. A. (2004). *Stress, cognition, and human performance: A literature review and conceptual framework* (NASA Tech. Memorandum No. 212824; p. 177). NASA Ames Research Center.

Stetson, C., Fiesta, M. P., & Eagleman, D. M. (2007). Does time really slow down during a frightening event? *PloS One*, *2*(12), e1295. https://doi.org/10.1371/journal.pone.0001295

Sulistyawati, K., Wickens, C. D., & Chui, Y. P. (2011). Prediction in situation awareness: Confidence bias and underlying cognitive abilities. *The International Journal of Aviation Psychology*, *21*(2), 153–174. https://doi.org/10.1080/10508414.2011.556492

Sweller, J. (1994). Cognitive load theory, learning difficulty, and instructional design. *Learning and Instruction*, *4*(4), 295–312. https://doi.org/10.1016/0959-4752(94)90003-5

Tedeschi, R. G., & Calhoun, L. G. (2004). Posttraumatic growth: Conceptual foundations and empirical evidence. *Psychological Inquiry*, *15*(1), 1–18.

The Globe and Mail. (2012, December 11). *Investigation finds stuck piston likely led to crash of CF-18 Hornet in air-show practice*. https://www.theglobeandmail.com/news/national/investigation-finds-stuck-piston-likely-led-to-crash-of-cf-18-hornet-in-air-show-practice/article6220481/

The New York Times. (2016, June 17). Explaining Greece's debt crisis. *The New York Times*. https://www.nytimes.com/interactive/2016/business/international/greece-debt-crisis-euro.html

Thoits, P. A. (2011). Mechanisms linking social ties and support to physical and mental health. *Journal of Health and Social Behavior*, *52*(2). https://doi.org/10.1177/0022146510395592

Thomson, D. R., Besner, D., & Smilek, D. (2015). A resource-control account of sustained attention: Evidence from mind-wandering and vigilance paradigms. *Perspectives on Psychological Science*, *10*(1), 82–96. https://doi.org/10.1177/1745691614556681

Thorpe, J. (2003). Fatalities and destroyed aircraft due to bird strikes, 1912–2002. *International Bird Strike Committee*, 28. https://ecirtam.net/autoblogs/autoblogs/frglobalvoicesonlineorg_0e319138ab63237c2d2aeff84b4cb506d936eab8/media/0b84ca6f.IBSC2620WPSA1.pdf

Tombu, M., & Jolicoeur, P. (2005). Testing the predictions of the central capacity sharing model. *Journal of Experimental Psychology. Human Perception and Performance*, *31*(4), 790–802. https://doi.org/10.1037/0096-1523.31.4.790

Tversky, A., & Kahneman, D. (1974). Judgment under uncertainty: Heuristics and biases. *Science (New York, N.Y.)*, *185*(4157), 1124–1131. https://doi.org/10.1126/science.185.4157.1124

U.S. Air Force. (2018). *Bird/wildlife aircraft strike hazard (BASH) management program* (No. 91–212; p. 52). U.S. Air Force. https://www.safety.af.mil/Portals/71/documents/Aviation/BASH/DAFI%2091-212%20Guidance%20Memorandum%208%20Jun%202022.pdf?ver=diwkouxyOQOYVp1RR8bWIg%3d%3d

U.S. Air Force. (2023). *Bird/wildlife aircraft strike hazard (BASH).* Bird/Wildlife Aircraft Strike Hazard (BASH). https://www.safety.af.mil/Divisions/Aviation-Safety-Division/BASH/

Vidulich, M., Wickens, C., Tsang, P., & Flach, J. (2010). Information processing in aviation. In *Human factors in aviation* (pp. 175–215). https://corescholar.libraries.wright.edu/psychology/289

Wickens, C. (2021). Attention: Theory, principles, models and applications. *International Journal of Human–Computer Interaction, 37*(5), 403–417. https://doi.org/10.1080/10447318.2021.1874741

Wickens, C. D. (2002). Situation awareness and workload in aviation. *Current Directions in Psychological Science, 11*(4), 128–133. https://doi.org/10.1111/1467-8721.00184

Wickens, C. D., & Alexander, A. L. (2009). Attentional tunneling and task management in synthetic vision displays. *The International Journal of Aviation Psychology, 19*(2), 182–199. https://doi.org/10.1080/10508410902766549

Wickens, C. D., Gutzwiller, R. S., & Santamaria, A. (2015). Discrete task switching in overload: A meta-analyses and a model. *International Journal of Human-Computer Studies, 79*, 79–84. https://doi.org/10.1016/j.ijhcs.2015.01.002

Wickens, C. D., Helton, W. S., Hollands, J. G., & Banbury, S. (2021). *Engineering psychology and human performance* (5th ed.). Routledge. https://doi.org/10.4324/9781003177616

Wickens, C. D., Santamaria, A., & Sebok, A. (2013). A computational model of task overload management and task switching. *Proceedings of the Human Factors and Ergonomics Society Annual Meeting, 57*(1), 763–767. https://doi.org/10.1177/1541931213571167

Wickens, C. D., & Yeh, M. (2018). *Display compellingness: A literature review* (DOT/FAA/AM-19/13; p. 18). United States. Department of Transportation. Federal Aviation Administration. Office of Aviation. Civil Aerospace Medical Institute. https://rosap.ntl.bts.gov/view/dot/56991

Wickens, C., Stokes, A., Barnett, B., & Hyman, F. (1993). *The effects of stress on pilot judgment in a MIDIS simulator* (pp. 271–292). Plenum Press. https://doi.org/10.1007/978-1-4757-6846-6_18

Wiegmann, D. A., & Shapell, S. A. (2016). *A human error approach to aviation accident analysis: The human factors analysis and classification system* (1st ed.). Routledge. https://www.taylorfrancis.com/books/mono/10.4324/9781315263878/human-error-approach-aviation-accident-analysis-douglas-wiegmann-scott-shappell

Wiggins, M., Stevens, C., Howard, A., Henley, I., & O'hare, D. (2002). Expert, intermediate and novice performance during simulated pre-flight decision-making. *Australian Journal of Psychology, 54*(3), 162–167. https://doi.org/10.1080/00049530412331312744

Wittmann, M., & Van Wassenhove, V. (2009). The experience of time: Neural mechanisms and the interplay of emotion, cognition and embodiment. *Philosophical Transactions of the Royal Society of London. Series B, Biological Sciences, 364*(1525), 1809–1813. https://doi.org/10.1098/rstb.2009.0025

Wright, N. A., & Haynes, S. R. (2009). Ejection decision-making: Can pilots accurately predict aircraft impact conditions? *Aviation, Space, and Environmental Medicine, 80*(9), 832–839.

VIDEO LINKS

CF18 crash Lethbridge. (2014, July 24). YouTube. https://www.youtube.com/watch?v=0HDIxzSMp-0

Flight demonstration of F-16 "Zeus" at Thessaloniki-28/10/2012 (English Translation from Greek Title: Α ε ρ ο π ο ρ ι κ ή Ε π ί δ ε ι ξ η F-16 "Ζ ε υ ς" σ τ η Θ ε σ σ α λ ο ν ί κ η—28/10/2012). (2012, November 19). YouTube. https://www.youtube.com/watch?v=4WQfD8DEl8Q

Raw video of an aircraft crashing in Poole Harbour. (2021, September 5). YouTube. https://www.youtube.com/watch?v=PRtSGb-17O4

Split second eject MiG-29 crash. (2018, March 30). YouTube. https://www.youtube.com/watch?v=sLH_LesiwkE

7 The Evolution of Air Shows

From Thrilling Spectacle to Safe and Sustainable Exhibition

7.1 INTRODUCTION

Air shows have captivated the hearts and minds of aviation enthusiasts and the wider public for over a century. Transitioning from the early days of daring barnstorming pilots performing in open cockpit biplanes to the contemporary era of precision aerobatics and cutting-edge displays featuring jet packs and drones, air shows have experienced remarkable transformations, entertaining millions of spectators around the globe. In this concluding chapter, we offer an insightful, forward-looking perspective on the exciting developments the air show industry may hold in store for the coming decades (see Figure 7.1).

Historically, air shows have adeptly adapted to the diverse needs of communities, and as we peer into the future, it is clear that they will continue to evolve while upholding a standard of excellence in every aspect. To remain relevant and enthralling, the industry must display agility in adjusting to the unique demands of different markets, demographics, communities, and nations. By doing so, air shows will not only endure as an exciting form of entertainment but also serve as a showcase for the latest breakthroughs in aerospace technology.

Nevertheless, securing support and sponsorship is a crucial element for the future of the air show industry, ensuring its financial sustainability. This backing can originate from a range of sources, including corporations, communities, municipalities, and government entities.

The air show industry is poised for a promising future in light of advancements in safety regulations and sustainable practices. Air shows are set to evolve into exemplary community events by embracing increased diversity and inclusivity, fostering collaborations and international partnerships. Consequently, these gatherings will embody a more responsible, eco-friendly form of entertainment that appeals to people of all ages and backgrounds while maintaining a steadfast commitment to industry-wide excellence.

To thrive in the future, the air show industry must address the ongoing challenge of meeting the varied expectations of diverse audiences. With rapid technological advancements and shifting consumer preferences, organizers need to innovate and craft engaging experiences that appeal to attendees from all walks of life. In order

DOI: 10.1201/9781003431879-7

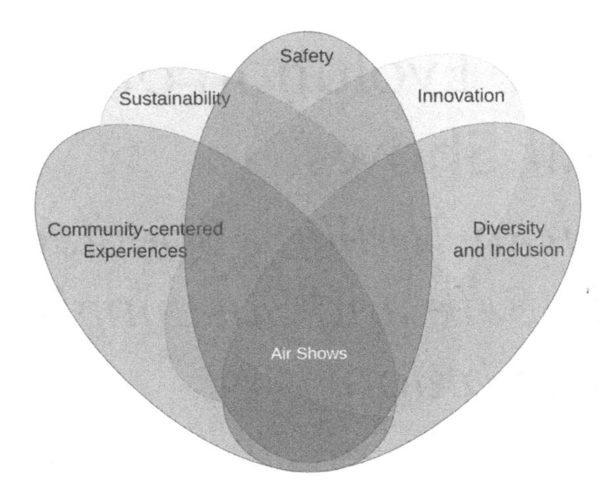

FIGURE 7.1 Air show industry's future emphasis areas.

to stay relevant and alluring, the future of air shows, particularly in Europe and the United States, will be influenced by a range of inventive trends and features. This chapter delves into these trends, emphasizing the necessary shifts in focus required to attract and engage a broad audience.

Ultimately, we expect air show performers to evolve to meet the requirements of an industry that is perpetually in flux. In this concluding chapter, we place emphasis on innovation, diversity, and sustainability, which will help them become versatile individuals possessing a wide range of skills. This adaptability will enable them to effectively respond to the ever-changing demands of the business, ensuring that air shows remain engaging and relevant for years to come.

7.2 SAFETY

The safety and well-being of air show attendees have always been and will continue to be of paramount importance to air show performers, organizers, and regulators in the future. Safety advancements are expected to focus on utilizing new technologies and methods for risk management. For instance, incorporating advanced data analytics and artificial intelligence in safety assessments could enable organizers and performers to identify and mitigate potential hazards more effectively. A practical application of this technology is using machine learning algorithms to forecast bird activity near the air show box, allowing for proactive measures and reducing the risk of bird strikes during air show performances.

7.3 SUSTAINABILITY

Future air shows should evolve in a manner that ensures the industry's environmental and financial sustainability. Organizers and performers should increasingly adopt eco-friendly practices, invest in cutting-edge technologies, and seek diverse funding sources to minimize their environmental footprint and maintain financial viability (Hagen, 2021; Raj & Musgrave, 2009). By embracing these sustainable approaches, the air show industry will ensure its longevity and continue to captivate audiences for generations to come, all while fostering a greater sense of responsibility toward our planet.

7.3.1 Environmentally Sustainable Practices

Air shows of the future must prioritize sustainability by embracing eco-friendly measures, such as harnessing alternative energy sources, employing electric tow vehicles, and utilizing solar-powered mobile command centers and briefing rooms. Investment in advanced, environmentally responsible technologies is crucial, including the development of electric ground support equipment, energy-efficient hangars, noise reduction technology, and comprehensive waste management and recycling programs. As a result, air shows can significantly decrease their environmental impact and contribute positively to the global effort against climate change.

7.3.1.1 Air Show Organizers

Air show sites and organizers throughout the world are increasingly concerned about environmental sustainability. Many people are using sustainable methods to address this problem, including trash reduction, energy-efficient lighting, and renewable energy sources. These initiatives reduce the adverse environmental effects of live events and meet the growing desire among attendees for eco-friendly experiences. In pursuing sustainability, we can expect continued innovation in alternative fuels and energy sources for aircraft, such as electric and hybrid planes.

Air show organizers also incorporate sustainable practices into their planning and execution processes. By implementing eco-friendly measures like recycling programs, compostable food service items, and carbon offset initiatives, they can work to reduce the overall environmental footprint of their events. Events like the Farnborough International Airshow in the United Kingdom exemplify this commitment to sustainability by incorporating waste reduction, energy conservation, and carbon offsetting strategies (Farnborough Airshow, 2023). By embracing sustainable practices and fostering environmental awareness, the air show industry can ensure a more responsible and eco-friendly future for spectators and the environment.

7.3.1.2 Air Show Performers

Sustainability is an essential aspect for air show performers as well. They can contribute to a more sustainable future by using electric or hybrid aircraft, incorporating environmentally friendly display routines, and promoting sustainable behavior among audiences. Additionally, performers can participate in sustainability initiatives implemented by air show organizers, such as being towed by electric vehicles to start points before starting engines to reduce the carbon footprints of taxiing and support sustainability through partnerships with environmental organizations or companies. By taking these steps, air show performers can be instrumental in raising environmental awareness and encouraging sustainable practices, helping to create a greener future for air shows.

7.3.2 Sustainable Financial Support and Sponsorship

Sponsorship has become increasingly crucial in air shows, as it enables organizers to secure funding from companies and organizations that want to promote their brands and support the aviation industry. To ensure financially sustainable support of the industry, sponsorship is an essential part of the future of air shows. For air show organizers, sponsorship is a critical source of funding that helps offset the high costs of organizing an air show. Air shows require significant investments in infrastructure, equipment, and talent, and sponsorship ensures that organizers can continue producing high-quality shows for years to come.

While there are challenges to be addressed, including ensuring that sponsors align with the values and mission of the air show, there are numerous benefits to both sponsors and air show organizers. As the aviation industry evolves, we expect changes in how sponsorship is approached, including technology, data analytics, and sustainability initiatives. By working together, sponsors and air show organizers can ensure that air shows continue to provide exciting and memorable experiences for attendees.

In addition to the trends previously mentioned, sponsors can explore numerous innovative approaches to support air shows. One such avenue is to promote research and development in aviation technology, such as electric or hybrid planes. For instance, in 2021, the U.K.'s Farnborough International Air show featured a demonstration flight by a zero-emission hydrogen fuel cell aircraft, the "Alpha Electro G2," produced by Slovenian aircraft manufacturer Pipistrel. This plane utilizes hydrogen fuel cells to power its electric motor, generating only water vapor as a byproduct.

Another progressive sponsorship initiative is supporting youth education, particularly in promoting aviation and science, technology, engineering, and mathematics (STEM) education. The U.S. Navy Blue Angels, for example, support STEM education programs that aim to inspire young people to pursue careers in aviation and engineering (Atkeison, 2018). By supporting such forward-thinking initiatives, sponsors can play a crucial role in the sustainable growth and development of the air show industry while simultaneously nurturing a passion for aviation in future generations.

7.3.3 Collaboration and International Partnerships

The air show industry is poised to experience a significant shift toward increased collaboration and international partnerships. As air show organizers and performers join forces to share knowledge, resources, and best practices, a cooperative atmosphere will be cultivated, leading to the development of innovative aerial displays and safety technologies. This collaborative spirit will ultimately elevate the standards and quality of air shows on a global scale.

Organizations like the International Council of Air Shows (ICAS) and the European Airshow Council (EAC) are already facilitating such cooperation by hosting annual conventions that assemble air show professionals from around the world. These gatherings provide a platform for networking, learning, and collaborating on strategies to enhance and advance the industry. By fostering these international connections, the air show community can continue to push the boundaries of what is possible in aerial entertainment, creating memorable and safe experiences for audiences worldwide.

Furthermore, efforts and initiatives can be made to popularize and enhance air shows in developing countries across sub-Saharan Africa, Southeast Asia, and South America, with the goal of tapping into the potential of emerging aviation powerhouses like Mexico, Brazil, Nigeria, Ghana, and Ethiopia. To further promote air shows in developing countries, the following recommendations can be considered:

- Collaboration with local aviation authorities: Engage with local aviation authorities and governments to develop supportive policies and regulations that facilitate the organization of air shows and foster a safe and thriving environment for aviation events.
- Partner with local airlines and aviation companies: Collaborate with local airlines, flight schools, and aviation companies to showcase their aircraft, services, and products, creating mutually beneficial partnerships that bolster the regional aviation industry.
- Establish mentorship programs: Encourage established air show organizations from developed countries to mentor and support emerging air show organizers in developing regions, sharing best practices, expertise, and resources.
- Develop regional air show circuits: Create regional air show circuits that include multiple countries, fostering cross-border collaboration and increasing visibility for these events in the targeted regions.
- Promote aviation education and outreach: Organize educational programs and community outreach initiatives alongside air shows to inspire young people to pursue careers in aviation and raise awareness about the industry's importance.
- Encourage cultural exchange: Showcase the unique cultural aspects of each host country, blending aviation and cultural performances to create distinctive and memorable air shows that celebrate local heritage and traditions.
- Leverage media and digital platforms: Utilize various media channels, including social media and live streaming, to promote and broadcast air shows, expanding their reach and impact on a global scale.

By implementing these recommendations and fostering international cooperation through organizations like ICAS and EAC, air shows in developing countries can gain momentum and contribute to the growth and development of the global aviation industry.

7.4 INNOVATION

Innovation in the form of technology will play a key role, as it always did, in the evolution of air shows in the future. Technologies such as virtual and augmented reality (VR/AR), holographic displays, artificial intelligence (AI), and unmanned aerial vehicles (UAVs). Concepts such as Urban Air Mobility (UAM) and effective social media integration into air show communication efforts will be part of this evolution.

7.4.1 Virtual and Augmented Reality

One of the most significant trends shaping the future of air shows is the integration of virtual and augmented reality (VR/AR) technologies. In virtual reality (AR), a user's perception of reality is based on additional computer-generated information within the data collected from real life and enhances their perception of reality (Ma et al., 2016). For example, in air show displays, VR can be used to create a simulation of a complex flight profile from the cockpit view with essential landmarks and features of the aerodrome superimposed on a real-life view.

On the contrary, augmented reality (AR) is an interactive experience for a user in which the real world and computer-generated content are combined and is defined as a system that incorporates three basic features: a combination of real and virtual worlds, real-time interaction, and accurate 3D registration of virtual and real objects (Wu et al., 2013).

The sensory information overlay in AR can be constructive (i.e., additive to the natural environment) or destructive (i.e., masking of the natural environment). This experience is seamlessly interwoven with the physical world such that it is perceived as an immersive aspect of the real environment (Rosenberg, 2022). AR differs from VR in the sense that in AR, part of the surrounding environment is "real," and AR is just adding layers of virtual objects to the real environment. On the other hand, in VR, the surrounding environment is completely virtual and computer generated (Carmigniani et al., 2011).

These technologies can also create immersive and interactive experiences for attendees, allowing them to engage with the performances in new and exciting ways and interact with aircraft displays in new ways. For example, such technologies could allow spectators, in real time, to virtually "ride along" with pilots during performances or view the show from unique perspectives, adding a new dimension to the air show experience. This trend is expected to enhance the overall experience for air show attendees and increase the popularity and accessibility of air shows worldwide.

7.4.2 3D Sound and Holographic Displays

As audio and visual technology continues to advance, the air show industry is poised to benefit from more immersive sound experiences and holographic displays. These cutting-edge innovations have the potential to dramatically enhance the overall experience for attendees, creating a highly engaging and lifelike atmosphere. By incorporating state-of-the-art sound systems, air shows can deliver richer and more realistic audio experiences, immersing spectators in the thrill of aerial performances.

Similarly, the integration of holographic[1] displays can revolutionize the way air shows are presented, offering visually stunning and dynamic simulations that captivate audiences. As the line between reality and virtual experiences becomes increasingly blurred, these technological advancements will usher in a new era of air shows, redefining the boundaries of aerial entertainment and captivating spectators in ways never before imagined.

There are examples of holography being used in other contexts, such as concerts, exhibitions, and trade shows. For instance, holograms have been used to bring deceased musicians back to life for performances, such as Tupac Shakur at Coachella in 2012 (Ganz, 2012) and Whitney Houston in her 2020 hologram tour (Whitney Houston Official Site, 2023). These instances demonstrate the potential of holography to create engaging and immersive experiences, which could be applied to air shows in the future.

7.4.3 Hybrid Events

The emergence of hybrid events, which blend in-person attendance with live-streaming options, is anticipated to significantly influence the future of air shows. This innovative format provides enhanced accessibility, allowing fans who may be unable to attend in person to enjoy live performances from the comfort of their homes. Hybrid events effectively eliminate barriers such as travel, cost, and mobility, making air shows more inclusive and widely available to a diverse audience.

Moreover, hybrid events offer organizers the opportunity to maximize their reach and revenue. By incorporating live-streaming options, air shows can expand their audience beyond the physical venue, attracting viewers from around the world. This increased exposure can not only drive ticket sales for future events but also generate additional revenue through virtual ticket sales and advertising.

The value of hybrid events became particularly evident during times of uncertainty, such as the COVID-19 pandemic. As public gatherings and large-scale events faced restrictions, hybrid formats provided a viable alternative for organizers to continue engaging fans and sustaining the air show industry. By adopting hybrid events, the air show community can adapt to changing circumstances while maintaining a connection with fans, ensuring the industry's long-term resilience and success.

7.4.4 Integration of Social Media

The integration of social media is anticipated to have a profound impact on the future of air show experiences, fostering real-time audience participation, content creation, and sharing. By incorporating social media platforms and digital technology into live events, organizers can captivate fans across generations, bridging the gap between age groups and fostering a sense of community. Social media also facilitates enduring connections between performers and their fans, allowing enthusiasts to follow their favorite acts, engage with them, and stay updated on their latest performances.

Moreover, integrating social media can lead to the creation of user-generated content, such as photos, videos, and live streams, which can be shared with a broader audience, further promoting the air show experience. This content can also serve as a powerful marketing tool, attracting new attendees and expanding the reach of the air show industry.

By embracing social media and digital technology, air shows can create unforgettable memories for spectators and cultivate a more interactive and immersive environment, ultimately enhancing the overall experience and ensuring the continued growth and success of the industry.

7.4.5 Shows with Small Unmanned Aerial Vehicles (UAVs)

Furthermore, the rise of small unmanned aerial vehicles (UAVs) could play a significant role in future air shows, potentially creating entirely new categories of aerial displays. Some displays feature formations of dozens or even hundreds of small UAVs (swarm displays), while others showcase the capabilities of cutting-edge UAV technologies like autonomous flight and advanced sensors. Swarm displays have demonstrated the ability to create stunning choreographed performances in the sky, offering new opportunities for creativity and innovation in air show programming.

As small UAVs become more sophisticated, air shows incorporate them into their performances, offering new and exciting experiences for spectators (see Figure 7.2). For example, the EAA (Experimental Aircraft Association) AirVenture in Oshkosh featured a drone light show in 2019, with over 100 small UAVs equipped with LED lights flown in a choreographed routine synchronized to music, which amazed audiences (PRWeb, 2019).

Considering that small and medium-sized UAVs might be a potential hazard to traditional aircraft and spectators, air show organizers and regulators should implement strict regulations and safety protocols for UAV displays in air shows. Firstly, UAV displays should be separated from traditional aircraft displays to reduce the chance of a collision. Additionally, cybersecurity constitutes a concern, as UAVs rely on complex software and communication systems, which can be vulnerable to hacking or other cyber threats. To address this, UAV operators and manufacturers must implement cybersecurity measures to guarantee that their aircraft and systems are secure and resistant to attacks, such as including user IDs or remote IDs to ensure that only authorized users at an air show can fly with minimal intrusions by rogue operators.

Communication problems might be an additional limitation in including UAVs in air shows performances. Most of them do not operate beyond the line of visual sight (BVLOS), and due to signal latency, as UAVs get further away from the operator, control inputs can be affected. That requires more sensitive and responsive transmitters and receivers, which can be disrupted by environmental factors such as interference or physical barriers. There should be well-established standard operating

FIGURE 7.2 Small UAVs show (drone show) at Cincinnati, USA. Copyright: Warren LeMay.

procedures (SOPs) for lost signals or lost communications and how UAVs will be recovered in such scenarios. If communication is lost, the UAV may not be able to receive essential commands from the operator, which can lead to a loss of control and potentially hazardous situations. In addition, using UAVs in air shows may require operators to communicate with multiple teams and individuals, which may increase the risk of miscommunication.

Another constraint involves the possibility of human error while operating UAVs. Since UAVs are typically controlled remotely, the operator might not possess complete knowledge of the aircraft's environment and potential hazards along its flight path. This limited situational awareness can heighten the chances of accidents and collisions, especially if the operator is unaware of the UAV's surroundings. Moreover, operators might need additional training when employing UAVs in air shows to reduce the probability of errors.

Lastly, limitations with camera feedback via links and other visual capture limitations can impair the situational awareness of the operator as their perception of the operational environment around the UAV is via a camera. Operating a UAV demands high levels of focus and mental acuity, with operators possibly facing fatigue if required to control the UAV for long durations.

7.4.6 Integration of Urban Air Mobility (UAM) Demonstrations

As the concept of urban air mobility[2] (UAM) continues to gain traction in the aviation industry, air shows are poised to feature demonstrations and showcases of these cutting-edge technologies. Audiences can expect to witness innovative concepts such as eVTOL (electric vertical takeoff and landing) aircraft, air taxis, and personal air vehicles on display, offering a glimpse into the future of urban transportation.

These showcases will not only introduce the public to new advancements in aviation technology but also serve as a platform for developers and manufacturers to demonstrate their latest achievements. By incorporating UAM technologies into air shows, organizers can generate excitement, foster public acceptance, and stimulate interest in the adoption of these emerging transportation solutions.

Furthermore, air shows that feature UAM technologies can inspire new generations of engineers, pilots, and entrepreneurs to pursue careers in the rapidly evolving field of urban aviation. Incorporating UAM technologies into air shows not only adds an exciting dimension to aerial entertainment but also highlights the industry's commitment to sustainable innovation, showcasing the potential for aviation to contribute positively to urban living and environmental stewardship.

7.5 DIVERSITY AND INCLUSIVITY

As society becomes more diverse and inclusive, air shows are expected to follow suit. This includes curating lineups featuring performers and exhibits from various backgrounds and ensuring that the overall event experience is welcoming and accessible to people of all abilities, genders, and ethnicities. By embracing diversity and inclusivity, air show organizers can attract a wider audience and foster a sense of community among attendees.

7.5.1 Diverse Air Shows

As we envision the future of the air show industry, it is crucial to recognize the growing emphasis on fostering diversity and inclusion within the aviation community. Air shows have the potential to function as powerful platforms for underrepresented groups, such as women and minorities, enabling them to demonstrate their skills and accomplishments in the field. By showcasing the talents of these individuals, air shows can help dismantle barriers, challenge stereotypes, and create an environment where diverse perspectives are celebrated.

Organizations dedicated to promoting diversity in aviation are already making strides in this direction. For example, the Women in Aviation International (WAI) organization hosts an annual Girls in Aviation Day event that features air show performances and educational activities designed to inspire and empower young women to explore careers in aviation. Events like these not only provide role models for aspiring aviators but also raise awareness about the need for greater representation and inclusivity within the industry.

By actively promoting diversity and inclusion at air shows, the industry can foster a more equitable and dynamic community where individuals from all backgrounds have the opportunity to excel and contribute their unique talents. This inclusive approach will not only strengthen the air show industry but also ensure its continued growth and success by attracting a diverse range of participants and audiences from around the world.

7.5.2 Accessible and Inclusive Events

Accessibility and inclusivity play a vital role in appealing to diverse audiences, encompassing individuals from various age groups, abilities, and backgrounds. In the future, air show organizers may place a higher emphasis on designing accessible venues, taking into consideration the needs of all potential attendees. By incorporating features such as ramps for wheelchair users, hearing loops for the hearing impaired, and sensory-friendly spaces for individuals with sensory sensitivities, air shows can become more welcoming and accommodating for all.

In addition to enhancing physical accessibility, organizers can focus on curating diverse lineups that represent different communities and backgrounds. By showcasing performers and acts from various cultures, gender identities, and walks of life, air shows can promote a sense of inclusivity and belonging, ensuring that all attendees feel represented and valued.

Emphasizing accessibility and inclusion in air show planning would not only create a more inclusive environment but will also widen the attractiveness of these events, attracting a larger spectrum of attendees. By encouraging a sense of community and respect for the various aviation enthusiasts, this inclusive approach will help the air show industry develop and succeed overall.

7.6 COMMUNITY-CENTERED EXPERIENCES

Focusing on community-centered experiences can help air shows in the future resonate with audiences of all ages and backgrounds. This may involve incorporating local aviation clubs, businesses, and organizations into the air show experience. By celebrating local culture and fostering a sense of community pride, air shows can create a more meaningful and engaging experience for attendees.

7.6.1 Intergenerational Entertainment

To captivate a diverse audience spanning multiple age groups, air show organizers might focus on crafting intergenerational entertainment experiences. This can be accomplished by showcasing a mix of classic and contemporary aircraft exhibits, effectively striking a balance between nostalgia and innovation. Additionally, incorporating activities and experiences specifically tailored to different age groups can further enhance the appeal of air shows for a broader demographic.

By offering a varied lineup and an assortment of experiences, air shows can effectively engage attendees of all ages, fostering intergenerational connections and shared enthusiasm for aviation. This approach not only strengthens the bond between older and younger generations but also facilitates the transfer of knowledge, passion, and appreciation for the industry.

Embracing intergenerational entertainment in air shows ensures the sustained growth and longevity of the industry by maintaining the interest of long-time enthusiasts while simultaneously captivating the imaginations of younger generations. This strategy cultivates a sense of continuity, preserving the rich history of aviation while inspiring the next generation of aviators, engineers, and spectators to carry the torch forward.

7.6.2 Educational and Outreach Programs

There is likely to be a growing emphasis on educational and outreach programs at air shows aimed at inspiring the next generation of pilots, engineers, and aviation enthusiasts. By engaging with young audiences and promoting STEM (science, technology, engineering, and mathematics) education, air shows can contribute to fostering a more diverse and inclusive industry.

For example, the Experimental Aircraft Association (EAA) AirVenture Oshkosh event in Wisconsin, United States, features a dedicated "KidVenture" area that offers a prime example of this approach. This specially designed space provides hands-on learning experiences for children and teenagers, sparking their interest in aviation and nurturing their passion for becoming future industry professionals (EAA, n.d.). Moreover, the organization of Black Aerospace Professionals (OBAP) has the Aviation Career Exposition camps where young people and kids are introduced to the world of aviation. Also, some Universities, such as the University of North Dakota (UND), hold annual Aviation Clinics for K–12 children to introduce them to aviation.

Incorporating educational and outreach initiatives into air shows not only enriches the overall event experience but also serves as an investment in the future of the aviation industry. By introducing young minds to the wonders of flight and the potential career paths in aviation, these programs can cultivate a new generation of innovators and enthusiasts who will drive the industry forward and ensure its continued success.

7.7 THE FUTURE AIR SHOW PERFORMER: EMBRACING INNOVATION, DIVERSITY, AND SUSTAINABILITY

The future air show performer will be a multifaceted individual, skilled not only in piloting and aerial acrobatics but also in embracing advanced technologies and adapting to the industry's changing landscape. As the air show industry evolves, performers must stay up to date with the latest innovations in aerospace, including electric and hybrid propulsion systems, autonomous flight capabilities, and cutting-edge aircraft designs.

Future air show performers must also be adept at engaging with diverse audiences. The industry aims to become more inclusive and accessible to people of all ages, backgrounds, and abilities. This may involve developing unique performance routines that cater to various interests, utilizing multimedia elements to enhance the spectator experience, and actively engaging with fans through social media and other digital platforms.

In addition, performers must prioritize safety and environmental sustainability in their acts. This may involve incorporating eco-friendly practices into their routines, such as reducing fuel consumption and emissions, utilizing biofuels or electric-powered aircraft, and adhering to strict safety protocols to minimize risks to both themselves and spectators.

Collaboration and partnerships will also be crucial for future air show performers. Working closely with other pilots, industry professionals, and international partners will enable performers to develop innovative routines, share knowledge and resources, and raise the overall standard of air show performances.

Moreover, this vision of the future air show performer is made more attainable with the availability of comprehensive aerobatics training in civilian flight schools. Much like their counterparts in the military, these performers can receive specialized training early in their careers, giving them a solid foundation on which to build their unique artistry.

All things considered, the future air show performer will be a versatile individual with a diverse skill set, able to adapt to the industry's ever-changing demands. Embracing advanced technologies, conducting specialized aerobatics flight training, engaging with diverse audiences, prioritizing safety and sustainability, and fostering collaboration will be essential attributes for performers as they strive to entertain and inspire future generations of aviation enthusiasts. Their spectacular feats of flight will ignite a passion for aviation in new generations, encouraging them to reach for the skies and dream and do big.

7.7.1 Innovative Performances

Air show performers have a significant role in shaping the future of air shows by continually developing innovative and thrilling performances. For example, the Breitling Jet Team, a renowned civilian aerobatic display team, gained considerable praise for its precision formation flying and groundbreaking aerial maneuvers. By constantly refining their routines, performers like the Breitling Jet Team captivate audiences while inspiring the next generation of pilots and aviation enthusiasts.

In the future, performers could explore more advanced technologies and techniques to create even more dynamic and immersive experiences for air show attendees. This may include collaborations with cutting-edge aerospace companies to incorporate new aircraft designs or propulsion systems into their routines, further pushing the boundaries of what is possible in aerial performances.

Another groundbreaking example is the Red Bull Air Race, which has transformed the concept of air shows by introducing high-speed, low-altitude racing through a series of inflatable pylons. This innovative format has provided a unique spectator experience while also demonstrating the potential for new types of aviation competitions. By continually exploring fresh ideas like these, air shows can attract more diverse audiences and generate interest in aviation as a sport beyond traditional air displays.

To sustain this momentum, air shows of the future could consider incorporating even more competitive elements, such as team-based challenges, cutting-edge aircraft technology, and interactive spectator involvement. By integrating these elements, air shows can continue to evolve and offer dynamic, engaging experiences that attract a wide range of attendees and maintain the industry's growth and vitality.

In the future, air show organizers and performers could revolutionize the industry by offering bespoke performances that incorporate real-time audience engagement, allowing spectators to decide specific maneuvers for performers to execute. This interactive approach would create an unprecedented level of excitement and personal investment for the audience as they become an integral part of the performance itself.

To achieve this, organizers could utilize digital platforms and apps that enable spectators to vote on a selection of aerial stunts, with the most popular choices being performed by the pilots. This real-time voting system could be facilitated through smartphones, tablets, or dedicated voting devices provided at the event. The chosen maneuvers would then be relayed to the pilots, who would seamlessly incorporate them into their routines, thrilling the crowd with their responsiveness and skill.

To ensure safety and feasibility, a predetermined set of maneuvers, approved by both the pilots and aviation authorities, could be presented as options for the audience to choose from. This approach would maintain the necessary safety standards while providing a unique, customizable experience for the spectators.

By offering this level of engagement, air shows could create a more immersive and memorable experience for attendees, fostering a stronger connection between performers and their fans. This pioneering approach would not only elevate the entertainment value of air shows but also contribute to the growth and sustained interest in the industry, attracting a wider audience and encouraging repeat attendance.

7.7.2 Advancements in Technology

Air show performers will also play a key role in advancing aviation technology. For instance, electric aircraft development has gained momentum in recent years, with several companies working on cutting-edge electric propulsion systems. Air show performers need to continue embracing this new technology by incorporating electric aircraft into their performances, showcasing the potential for a cleaner, quieter, and more sustainable future in aviation.

Furthermore, developments in cockpit display technology can simplify primary flight displays and alerting systems, improving pilots' situational awareness. Automation systems, such as those demonstrated by Cessna and Piper (Boatman, 2020), can take control of the aircraft and secure a safe landing in cases of pilot incapacitation due to high G maneuvers or medical events. Innovations like Cirrus' parachute pack for immobilized aircraft (Cirrus Approach, 2023) offer safety solutions for smaller display aircraft like the Extra 300. Furthermore, high-fidelity goggles or spectacles with Head Mounted Display (Bayer et al., 2009; Chaparro et al., 2023; Ernst et al., 2021; Wells and Haas, 2020) capabilities can present crucial flight information within the pilot's conformal view, minimizing head-down movements and enhancing safety during performances.

Additionally, advancements in augmented reality (AR) and virtual reality (VR) have opened up new possibilities for air show experiences. Performers could leverage these technologies to create immersive experiences that allow spectators to enjoy air shows in ways never before possible. Examples include virtual cockpit tours, interactive displays, and live streaming of air shows with multiangle camera perspectives.

7.7.3 New Air Show View: Shifting to a Culture of Excellence as the Core Ingredient for Safety

The air show industry is poised to embrace a new view on safety, focusing on cultivating a culture of excellence as the fundamental aspect of ensuring safety in performances. By prioritizing excellence in every aspect of air shows, the industry can inspire new generations of pilots and aviation enthusiasts while creating a positive and proactive approach to safety.

This shift to a culture of excellence will involve a comprehensive and holistic approach, encompassing not only the technical skills of pilots but also their mental and physical well-being. Performers will be encouraged to continuously refine their skills, attend workshops, and participate in training sessions that promote excellence in all aspects of their performance, including communication, teamwork, and decision-making.

Moreover, there is room for industry-academia collaboration that will collect data and conduct research on pilot performance during airshows. Such physiological data and psychological studies can help refine the existing understanding of the challenges faced by air show performers. It will also influence policy and practices on safety and human performance standards.

Air show organizers will also play a crucial role in fostering this culture of excellence. By setting high standards for their events, they can encourage performers to strive for perfection and push the boundaries of what is possible in aerial displays. Moreover, organizers can provide resources and opportunities for performers to engage in ongoing professional development, further enhancing their skills and expertise.

An essential part of this new culture of excellence will be the sharing of knowledge and best practices among performers, organizers, and the wider air show community. By promoting open communication and collaboration, the industry can ensure that everyone benefits from the collective wisdom and experiences of their peers, resulting in safer and more impressive performances.

The pursuit of excellence in the air show industry will undoubtedly have a significant impact on audience experiences. Organizers are encouraged to create engaging and accessible events that captivate spectators, fostering a sense of wonder and excitement about the world of aviation. To achieve this, a diverse range of experiences should be offered, including educational programs, interactive exhibits, and state-of-the-art technological displays. Such experiences can inspire and educate people of all ages and backgrounds, drawing them closer to the world of aviation.

In addition to the unforgettable performances and attractions, creating a sense of community and pride among attendees is crucial. By offering merchandise such as T-shirts, patches, and stickers that celebrate air show excellence, spectators can feel a sense of belonging and pride in being part of this remarkable showcase of human achievement. These tangible mementos serve as lasting reminders of their positive experiences and can help to create a loyal fan base eager to return and share their passion for aviation with others.

Ultimately, the new view on safety in the air show performers, and the industry as a whole, will center around the pursuit of excellence in all aspects of performances and organization. This shift toward a culture of excellence will not only enhance safety but will also serve to inspire new generations and foster a positive, vibrant atmosphere within the air show community. By striving for excellence, the industry can achieve both the pinnacle of aerial displays and the highest standards of safety, ensuring a bright future for air shows worldwide.

7.7.4 Environmental Sustainability

As the global awareness of environmental concerns related to aviation increases, air show performers are progressively adopting eco-friendly practices to mitigate their environmental impact. One notable example, as mentioned earlier, is the shift toward using electric aircraft, which produce significantly fewer emissions than traditional fuel-powered planes.

In addition to adopting cleaner aircraft technology, air show performers also need to embrace other sustainable practices. Working closely with organizers, they could implement strategies such as carbon offsetting, where emissions produced during air shows are offset by investing in projects that reduce greenhouse gas emissions elsewhere, effectively balancing out their environmental impact.

Another essential component of sustainable air shows is the adoption of sustainable aviation fuels (SAFs). These alternative fuels, often derived from renewable resources like waste oils, agricultural residues, or algae, can substantially reduce the carbon footprint of aviation activities. By utilizing SAFs, air show performers can demonstrate their commitment to environmental stewardship while still delivering thrilling performances.

Continuing with their commitment to sustainability, air show performers may also focus on minimizing waste generated during events. Implementing waste reduction strategies such as using reusable or biodegradable food service items and promoting responsible waste management among attendees can contribute to greener air shows.

Educational programs and activities centered around environmental conservation could also be integrated into air show events. By engaging with audiences, especially younger generations, air show performers can raise awareness about the importance of sustainable practices in aviation and encourage the adoption of eco-friendly behaviors among attendees.

Partnerships with environmentally focused organizations, nonprofits, and sponsors can also play a crucial role in promoting sustainability at air shows. By collaborating with like-minded stakeholders, air show performers can access resources, expertise, and funding to support their green initiatives, further solidifying their commitment to environmental responsibility.

In addition, eco-friendly travel options can play a significant role in minimizing the overall carbon footprint of air show events. By encouraging attendees to use sustainable modes of transportation, air show performers can demonstrate their commitment to environmental sustainability and reduce the impact of their events on the planet. Air show performers can promote carpooling among attendees, enabling people to share rides and reduce the number of vehicles on the road.

7.8 BIKING AT AIR SHOWS: DEMONSTRATION OF COMMITMENT TO SUSTAINABILITY AND WELL-BEING

Biking is indeed another eco-friendly travel option that can be promoted at air shows, and providing bicycles to air show performers for on-site movement is an excellent way to further emphasize the event's commitment to sustainability. By utilizing bicycles for transportation within the venue, performers can reduce their reliance on motorized vehicles, thereby decreasing fuel consumption and emissions.

Organizers can collaborate with local bike rental companies or invest in a fleet of bicycles for use by performers and staff during the event. Designated bike parking areas, maintenance stations, and clearly marked paths can be set up throughout the venue to ensure a safe and convenient biking experience for everyone involved.

In addition to benefiting the environment, promoting biking for air show performers can have several other advantages. For one, it can help reduce traffic congestion within the event area, allowing for smoother operations and quicker response times in case of emergencies. It also promotes physical fitness and well-being among the performers, who may appreciate the opportunity to stretch their legs and get some exercise between performances.

Furthermore, by visibly embracing sustainable practices such as biking, air show performers can inspire attendees to consider adopting similar eco-friendly habits in their daily lives. This can help raise awareness about the importance of sustainable transportation and contribute to a more environmentally conscious society.

Overall, providing bicycles to air show performers for on-site movement is a practical and eco-friendly initiative that can enhance the event's sustainability efforts while promoting a healthy lifestyle and setting a positive example for attendees.

To maintain the appeal of air shows in the face of growing environmental concerns, it is crucial for the industry to continuously adapt and innovate. By embracing sustainable practices, prioritizing eco-friendly technologies, and engaging with audiences on environmental matters, air show performers, along with organizers, can ensure that these awe-inspiring events continue to flourish while minimizing their impact on the planet. In doing so, they not only entertain and inspire but also contribute to a more sustainable future for aviation and the broader global community.

Air show performers play a crucial role in shaping the future of air shows by introducing innovative performances, embracing new technologies, prioritizing safety, and committing to environmental sustainability. As these performers continue to push the boundaries of aviation entertainment, they will undoubtedly inspire new generations of pilots and aviation enthusiasts, ensuring the longevity and vitality of air shows for years to come.

NOTES

1. A hologram is a three dimensional (3D) image created by recording the interference pattern generated by the interaction of laser light with an object. This technique is based on the principles of holography, a photographic process developed by Hungarian British physicist Dennis Gabor in 1947. Holography captures the amplitude and phase information of light waves reflected from an object, which can then be reconstructed to produce a 3D image that appears to have depth and can be viewed from different angles without the need for special glasses (Saxby & Zacharovas, 2015).

   Holographic technology has a diverse array of applications across various industries, including entertainment, advertising, art, medicine, education, and augmented/virtual reality. Holograms have been utilized in concerts to bring back deceased artists, in advertising for eye catching product displays, and in museums for immersive installations. In the medical field, holography is used for imaging and surgical planning, while holographic displays in educational settings facilitate interactive learning experiences. Lastly, holographic technology is crucial in advancing augmented and virtual reality devices, providing users with more realistic experiences.
2. Urban Air Mobility (UAM) is an emerging concept within the aviation industry that envisions the use of small, electric or hybrid electric aircraft to transport passengers and cargo within urban and suburban areas. These aircraft are designed to be more environmentally friendly, quieter, and efficient than traditional helicopters or small aircraft. UAM aims to reduce ground congestion, decrease travel times, and increase transportation options for people in densely populated cities .

   The concept of UAM gained traction in recent years as advances in electric propulsion, battery technology, and autonomous systems have made the development of electric vertical takeoff and landing (eVTOL) aircraft more feasible. Several companies are working on prototypes and conducting test flights, aiming to commercialize UAM in the coming years.

   The successful implementation of UAM depends on overcoming a variety of challenges, including regulatory approval, airspace integration, infrastructure development, public acceptance, and safety concerns. Regulatory bodies like the Federal Aviation Administration (FAA) in the United States (Federal Aviation Administration, 2023) and the European Union Aviation Safety Agency (EASA) (EASA, 2021) in Europe are working on developing guidelines and standards for the safe operation and integration of UAM systems into existing airspace.

LIST OF REFERENCES

Atkeison, C. (2018, August 25). *U.S. Navy's Blue Angels inspire young adults in STEM careers*. Avgeekery.Com. https://avgeekery.com/u-s-navys-blue-angels-inspire-young-adults-in-stem-careers/

Bayer, M. M., Rash, C. E., & Brindle, J. H. (2009). *Introduction to helmet-mounted displays: U.S. Army Aviation*. https://doi.org/10.1037/e614362011-004

Boatman, J. (2020, March 5). Garmin's autoland gets flight tested. *Flying Magazine*. https://www.flyingmag.com/garmin-autoland-flight-tested/

Carmigniani, J., Furht, B., Anisetti, M., Ceravolo, P., Damiani, E., & Ivkovic, M. (2011). Augmented reality technologies, systems and applications. *Multimedia Tools and Applications, 51*(1), 341–377. https://doi.org/10.1007/s11042-010-0660-6

Chaparro, A., Miranda, A., & Grubb, J. (2023). Chapter 13 – Aviation displays: Design for automation and new display formats. In J. R. Keebler, E. H. Lazzara, K. A. Wilson, & E. L. Blickensderfer (Eds.), *Human factors in aviation and aerospace (Third Edition)* (pp. 341–371). Academic Press. https://doi.org/10.1016/B978-0-12-420139-2.00014-9

Cirrus Approach. (2023). *CAPS training*. Cirrus Approach. https://www.cirrusapproach.com/caps-training/

EASA. (2021, May 19). *Urban Air Mobility (UAM)*. EASA. https://www.easa.europa.eu/en/domains/urban-air-mobility-uam

Ernst, J., Ebrecht, L., & Korn, B. (2021). Virtual cockpit instruments—How head-worn displays can enhance the obstacle awareness of helicopter pilots. *IEEE Aerospace and Electronic Systems Magazine*, *36*(4), 18–34. https://doi.org/10.1109/MAES.2021.3052304

Farnborough Airshow. (2023). *Farnborough airshow: Sustainability*. Farnborough Airshow. https://www.farnboroughairshow.com/the-show/industry-themes/sustainability/

Federal Aviation Administration. (2023). *Urban air mobility and advanced air mobility*. FAA. https://www.faa.gov/uas/advanced_operations/urban_air_mobility

Ganz, J. (2012, April 17). How that Tupac hologram at Coachella worked. *NPR*. https://www.npr.org/sections/therecord/2012/04/17/150820261/how-that-tupac-hologram-at-coachella-worked

Hagen, D. (2020). *Sustainable event management: New perspectives for the meeting industry through innovation and digitalisation?*

Ma, M., Jain, L. C., & Anderson, P. (2016). *Virtual, augmented reality and serious games for healthcare* (1st ed.). Springer Publishing Company, Incorporated.

PRWeb. (2019, February 20). *Drone light show returns to illuminate night sky at EAA AirVenture Oshkosh 2019*. PRWeb. https://www.prweb.com/releases/drone_light_show_returns_to_illuminate_night_sky_at_eaa_airventure_oshkosh_2019/prweb16115195.htm

Raj, R., & Musgrave, J. (Eds.). (2009). *Event management and sustainability*. CABI Publishing.

Rosenberg, L. B. (2022). Augmented reality: Reflections at thirty years. In K. Arai (Ed.), *Proceedings of the Future Technologies Conference (FTC) 2021, volume 1* (pp. 1–11). Springer International Publishing. https://doi.org/10.1007/978-3-030-89906-6_1

Saxby, G., & Zacharovas, S. (2015). *Practical holography* (4th ed.). CRC Press. https://www.routledge.com/Practical-Holography/Saxby-Zacharovas/p/book/9781482251579

Whitney Houston Official Site. (2023). *An evening with Whitney: The Whitney Houston hologram concert*. Whitney Houston Official Site. https://www.whitneyhouston.com/tour/an-evening-with-whitney-the-whitney-houston-hologram-tour/

Wells, M. J., & Haas, M. (1992). The human factors of helmet-mounted displays and sights. In *Electro-optical displays* (p. 44). CRC Press.

Wu, H. K., Lee, S. W. Y., Chang, H. Y., & Liang, J. C. (2013). Current status, opportunities and challenges of augmented reality in education. *Computers and Education, 62*, 41–49. https://doi.org/10.1016/j.compedu.2012.10.024

8 Concluding Remarks

As we conclude this book, it is crucial to recognize the profound impact of air shows on our communities. Since their inception, air show performers have been regarded as legendary figures, embodying the spirit of mythological heroes such as Icarus.

The myth of Icarus can teach air show performers valuable lessons about ambition, risk assessment, and the importance of adhering to safety guidelines. In the myth, Icarus and his father, Daedalus, crafted wings made of feathers and wax to escape from the Island of Crete. Daedalus warned Icarus not to fly too close to the sun or the sea. However, Icarus, caught up in the exhilaration of flight, ignored his father's advice and flew too close to the sun. The heat melted the wax, causing Icarus to fall into the sea and drown.

The relevance of the myth of Icarus to modern air show performers lies in the cautionary tale it provides. Like Icarus, air show performers may be captivated by the thrill of flight and the desire to push boundaries. However, the myth reminds them to respect the limitations of their aircraft and their own abilities. It teaches them the importance of heeding expert advice, maintaining situational awareness, and prioritizing safety over spectacle.

By internalizing the lessons from the myth of Icarus, air show performers can strike a balance between showcasing their skills and ensuring the safety of themselves, their fellow performers, and the spectators. It encourages them to recognize the potential dangers associated with pushing the limits and managing risks responsibly, ultimately leading to a safer and more sustainable air show industry.

In concluding this book, we pay homage to air show performers who have paid the ultimate price for the freedom of flight and commemorate the spectators and bystanders who have tragically lost their lives during air show events, mesmerized by the accomplishments of human innovation. Drawing inspiration from the myth of Icarus, we hope that future incidents will be prevented. This book we envision contributes to fostering wiser, safer, and more exceptional air show performers who prioritize risk assessment, accident prevention, and mindful management of safety factors.

DOI: 10.1201/9781003431879-8

Glossary

Manolis Karachalios and Daniel Kwasi Adjekum

Numerous definitions exist throughout international aviation authorities involving air show performers and air bosses, which may confuse the reader. However, for this book, "air show performers" will be defined as all pilots/operators flying any aircraft or unmanned air vehicle (UAV) who perform aerobatics–solo or formation, or dynamic maneuvering – solo or formation. Then, an "air boss" will be mentioned as the person in charge of the flight operations at an aviation event.

In addition, the following definitions will be referenced under the UK CAA's CAP 403[1]:

Crowd line	The line delineating the closest edge of any area, including car park(s), accessible to spectators concerning the display area/display line.
Display line	A line defining the track along which displaying aircraft may operate.
Flying display	Any flying activity deliberately performed for the purpose of providing an exhibition or entertainment at an advertised event open to the public.
Flying control committee (FCC)	A group of suitably experienced persons assembled to assist the FDD in the safety management of a flying display.
Flying display director (FDD)	The person responsible to the authorities for the safe conduct of a flying display.
Show center or display datum	The display datum is the point upon which individual displays are based and is normally the center point of the crowd.
Spectator area	An area specifically designated for spectators by the event organizer or FDD and approved by the FDD for flying display safety purposes which includes all areas to which spectators have access during the flying display.
Minimum aerobatic height	The minimum height above which the aircraft must be capable of complete recovery from an aerobatic maneuver. This will be the most restrictive of the following: • The minimum aerobatic height specified in the Permission. • The minimum aerobatic height quoted on the relevant pilot's display authorization (in relation to the aircraft category being flown); or • The minimum aerobatic height imposed by the FDD.

NOTE

1. UK Civil Aviation Authority. (2023). *CAP 403: Flying Displays and Special Events: Safety and Administrative Requirements and Guidance*. UK Civil Aviation Authority. https://publicapps.caa.co.uk/docs/33/CAP403%20Edition%2020.pdf

DOI: 10.1201/9781003431879-9

Index

A

Abbott Robert 21
accessibility 210, 211, 215
Aero South Africa air show 26
aerobatic team 32, 34, 52, 67
  AeroSuperBatics 11, 195–198
  Airborne Pyrotechnics 9
  Blades aerobatic team 25
  Blue Angels, U.S. Navy 5–7, 30, 31, 34, 48, 100, 109, 143, 161, 162, 208,
  Frecce Tricolori, Italian AF 104
  Fursal Al Emarat, United Arab Emirates AF 5
  Misty Blues Skydiving Team 24
  Patriot Jet Team 101
  Patrouille de France, French A.S.F. 5, 25, 27
  Red Arrows, United Kingdom Royal Air Force 5, 25, 78, 79, 100, 101, 182–187
  Red Baron, Australia 26
  Royal Jordanian Falcons 25
  Sky Hawk Parachute Demonstration Team 181
  Snowbirds, Royal Canadian Air Force 5, 48, 100, 109
  The Five Blackbirds 23, 24
  Thunderbirds, United States Air Force 27, 30, 34, 48, 100, 109
  Turkish Stars, Turkish Air Force 52, 193
  Wefly! Team 19
aerobatic competency evaluator (ACE) 89, 107
African American 20, 21, 23–25
anti-G straining maneuver (AGSM) 132
air boss 1, 13–15, 29, 39, 57–59, 61, 64–67, 70, 73, 80, 82, 85, 89, 108, 109, 112, 114, 115, 119, 126, 155, 157, 158, 190, 225
air races
  Cleveland National Air Races 34
  Reno Air Races 48, 49, 90
air show organizers 16, 31, 47, 52, 57, 70, 79, 80, 85, 114, 115, 139, 194, 195, 207–209, 212, 214, 215, 218, 219
air show safety briefing 60, 154, 157
air traffic control (ATC) 57, 67, 176
almost loss of consciousness (A-LOC) 132, 133
antidote 119, 130, 131
anxiety 31, 32, 59, 64, 118, 123, 133, 134, 144, 145, 147, 152, 159, 161
Argyropoulos Emmanouil 29, 39
arousal 126, 193, 196
artificial intelligence (AI) 206, 210
as low as reasonably practicable (ALARP) 45
asymmetrical thrust 191
Athens Flying Week (AFW) International Air Show 9, 19, 78, 79
attention tunneling 171, 173
augmented reality (AR) 210, 219
Auriol Jacqueline 22
automaticity 144, 175, 180
Avalon Airshow 26

B

B–25 Mitchell bomber 8
B-52 Fortress 50, 51, 119
Barker Des 48, 49, 66, 76, 119, 123
Barnes Pancho 17, 37
barnstormers 4, 30, 32, 33, 36–38
Bastié Maryse 20, 22, 23, 29
Beachy Linkoln 29
Bews Brian 177–181
bias 30–32, 120–126
  nchoring bias 120–123, 126, 138, 184, 197
  confirmation bias 30, 120–125, 138, 171–173, 184
  expectation bias 188, 190
  groupthink bias 31
  overconfidence / illusory superiority bias 30, 32, 51, 62, 120–125, 128, 138, 140, 171–173, 189, 190
  Plan continuation bias 30, 120, 121, 125, 137, 171–173
  social facilitation bias 65, 120–123, 138, 170–173
biking 221
bird strike 49, 52, 59, 65, 82, 87, 191–195, 206
bird / wildlife aircraft strike hazard (BASH) 68, 194
Blériot Louis 35, 36
Bothelin Jacques i, 4
Boucher Hélène 22
Bournemouth Air Festival 186, 195
Breitling Jet Team 31, 217
British Air Display Association (BADA) 16, 17
British Gliding Association (BGA) 9
British Hang Gliding and Paragliding Association 10
British Skydiving 10

buffers 68, 174
safety buffer 60, 174
time buffer155

C

CF-188/ CF-18 Hornet 177–182
Carlton Bob 8
central bottleneck effect 170, 171, 173, 174
chair flying 152
Chipmunk DHC-1 aircraft 50
Civil Aviation Authority (CAA) 15, 16, 31, 68, 76, 98
Clark Julie 31
Cockburn George 36
Cognitive
cognitive capacity 136
cognitive overload 193
Coleman Bessie 17, 20–22, 37
Commemorative Air Force (CAF) 1, 7, 50
communication
communication errors 188, 189
communication protocols 81, 197, 198
concealed hazardous attitudes 127–130
conditioning 134, 184
consistency 152, 156
continuous enhancements 99, 105, 110
continuous improvement 12, 99, 100, 111, 187
controlled flight into terrain (CFIT) 166
CounterMeasures dispensing system (CMDS) 170, 173
COVID-19 16, 68, 97, 211
crew resource management (CRM) 139, 190
culture
air show culture 96, 97, 100, 105, 106, 114,
change culture 105
discipline culture 108
excellence culture 100, 103, 105, 106
No-Blame culture 182
operational culture 105
resilience safety culture 99
resilient safety culture i, xii, xiii, 38, 39, 55, 69, 76, 97–99, 110–112, 152
safety culture xii, xiii, 1, 2, 38, 39, 45, 55, 69, 71, 73, 89–91, 97, 98, 105–108, 110, 113–115, 131, 152, 182, 186, 187
Cunningham Sean, Flight Lieutenant 186
Curtiss Glenn 30, 33, 35, 36,

D

Daredevils 29,
decision making 30, 33, 53, 54, 60, 64, 69, 72, 75, 85, 119–121, 123–126, 131–136, 138, 140, 145–147, 149, 152, 159, 166, 171–174, 177, 180, 183–185, 188, 190, 193, 196, 197, 219
aviation decision making 46
ejection decision making 180
demographics 75, 76, 205, 215
disorganization 127, 130, 131
distraction xii,49, 51, 59,62, 110–112, 127, 130, 131, 136, 137, 139, 140, 143, 155, 157, 158, 159, 162, 176, 184
distractions in the cockpit 52, 171
self-induced distraction 167, 170, 176, 177
ditching 195, 198
diversity 1, 3, 17–19, 23–27, 35, 205, 206, 214, 216, 217
Dominguez Air Meet 33
drone racing league (DRL) 31
Dussia Jase 11

E

Egging Jon, Flight Lieutenant 186
eject/ejection 51, 52,177, 179–181, 183, 186, 192, 193
ejection decision 185
ejection seat 110, 177, 181, 185
ejection system 180, 193, 194
electric vertical takeoff and landing (eVTOL) 214, 222
emergency response 78, 181, 193, 194
emergency response planning 79, 191
emergency response team 78, 181, 182, 194, 198
energy management traps 175, 199
European Airshow Council (EAC) 14, 16, 17, 78, 109, 195, 208, 209
Experimental Aircraft Association (EAA) Airventure Oshkosh 216

F

F-16 Fighting Falcon 5, 6, 34, 112, 159, 160, 167, 168, 172, 199
F-18 Hornet 5, 34
Fairchild Air Force Base 51, 119
falcon turn maneuver 168, 169, 172, 174
Farman Henri 36
Farnborough International Airshow 27
fast jet solo display pilot safety workshop 112–115
Federal Aviation Administration (FAA) 13, 16, 29, 119, 127, 130, 192, 222
Fédération Aéronautique Internationale (FAI) 17
Fédération des Spectacles Aériens (FSA) 16
field observation 66
flight into object (FIO) 50
flight into terrain (FIT) 50
flight path marker (FPM) 74, 75
flow 81, 146, 160
display flow 68, 148
mental flow 62

flying display director (FDD) 14, 15, 66, 68, 70, 71, 107, 119, 194
flying control committee (FCC) 13, 14, 47, 66, 68
focus 31, 34, 62, 82, 131, 133, 140, 146–150, 155–157, 161, 162, 189, 193, 213
  balanced focus 173
  collective focus 151
  mental focus 150, 151, 160
fear of missing out (FOMO) 32, 33
formation flying 3, 5, 8, 11, 27, 50, 56, 66, 75, 101, 104, 217
Formula 1 racing 111, 115
French Air and Space Force's Rafale Solo Display 143, 160

G

Gates Flying Circus 37
ground collision avoidance system (GCAS) 174, 176
get-there-itis 172
G-induced loss of consciousness (G-LOC) 50, 51, 62, 131, 132
G-induced vestibular dysfunction (GIVD) 132
Guyot Virginie 25

H

hands on throttle and stick (HOTAS) 170, 174
Hawker Hunter 50, 52, 132, 136, 138
Hawker Hurricane 49, 187–191
hazardous attitudes 30, 39, 50, 54, 105, 118, 119, 122, 126–131, 152, 157, 171
  concealed hazardous attitudes 127–131
head mounted display 218
head-up display (HUD) 74, 75, 169, 173, 175
heuristics 120, 171, 173, 174
high-reliability organization (HRO) 111, 115
Holland Rob 26
holographic displays 210, 222
Hoover Bob 76
human factors 30, 31, 50, 51, 59, 112, 184, 185
  human factors analysis and classification systems (HFACS) taxonomy 118
hybrid events 211

I

IMSAFE checklist 72
inclusivity 17, 19, 25, 26, 34, 205, 214, 215
International Civil Aviation Organization (ICAO) 46, 88, 105
International Council of Air Shows (ICAS) 14, 16, 29, 71, 100, 109, 208
innovation 3, 4, 12, 13, 26, 27, 32, 35, 79, 96, 100, 101, 206, 207, 210, 212, 214–216, 218, 224

K

Kapanina Svetlana 18
Kvochur Anatoly 191–194

L

lapses in attention 189, 190
Latham Hubert 36
Lefebvre Eugene 36
Lemordant Aude 103
Lethbridge International Air Show 177–182
Lindbergh Charles 37
loss of control in-flight (LOC-I) 49

M

maintaining vigilance 188–190
Malta International Air Show 25
Marvingt Marie 17
mechanical failure 33, 47–49, 52, 58
mental anchoring 149, 150
mental rehearsal 158, 159
mentorship 18, 76, 77, 114, 127, 209
Mfeka Mandisa 26
mid-air collision (MAC) 49, 50, 57, 82
Midnight Sun Air Show 25
MIG-29 191–194
mindfulness 39, 150–153, 156–158, 161, 162
  Mindfulness-based Mind Fitness Training (MMFT, or M-fit) 145
  mini-mindfulness 147–149
  synchronized mindfulness 150, 151
mitigation strategy 86, 87
Murphy Kirsty 25
MXS Aircraft 52
Myth of Icarus 224

N

National Gay Pilots Association (NGPA) 18
NF-5 Freedom Fighter 52
normalization of deviance 45, 89–91, 108

O

operational risk management 55, 58, 79, 80
  airshow operational risk management (AORM) 45, 80–87
Organization of Black Aerospace Professionals (OBAP) 18, 216
Oshinuga Anthony 24, 25
Otto the Helicopter 7
overconfidence 30, 32, 51, 62, 120–125, 128, 138, 140, 171–173, 189, 190
ownership 105, 109

P

P-51 Mustang 8, 49
PC-21 Pilatus 50, 51
Paris Air Show 26, 27, 32, 191–193
Paulhan Louis 30, 33
Pégoud Adolphe 30
Pietsch Kent 26
post-traumatic stress disorder (PTSD) 145
probability 45–47, 53, 81, 83, 85–87, 89, 213
psychological refractory period (PRP) effect 170, 171
push–pull effect 51, 132–134

Q

Quad City Airshow 52

R

Radius Manfred 8
Radom Air Show 50
RC Kavala Acro Team 11, 12
Red Bull
  Red Bull Air Force 10
  Red Bull Air Race 12, 31, 104, 111, 136, 217
  Red Bull X-Alps 11
  The Flying Bulls 103, 104
Reims Air Meet 33, 35, 36
resetting 148, 156
resilience 19, 20, 22, 27, 29, 33, 47, 67, 68, 82, 98, 99, 109, 112, 118, 134, 144, 147, 148–150, 166, 167, 178, 182, 186, 198, 211
ribbon cut 23, 102, 115, 155, 156
Royal International Air Tattoo (RIAT) 6, 8, 25, 39, 101, 179, 180, 183, 191
Royal Navy Black Cats 7, 8
risk 7, 13, 16, 29, 30–36, 38, 39, 45–62, 65–91, 96, 97, 100, 101, 104, 105, 107, 109–112, 114, 118, 119, 121–127, 130, 133, 135, 137–140, 147, 150–152, 154, 156–158, 161, 162, 166–168, 170, 171, 174, 178, 180–182, 185, 186, 189, 190, 192, 194, 195, 198, 199, 206, 213, 217, 224
  calculated risk 46, 167
  financial risk 53, 55, 56
  risk assessment matrix 59–61
  risk level 79, 81, 83, 85–87, 152
  risk management 1–3, 13–15, 17, 29, 31, 35, 45–48, 52, 54, 55, 57, 58, 62, 70–75, 77–82, 85, 88, 100, 101, 111, 114, 157, 166, 164, 166, 181, 185, 190, 193, 197, 198, 206
  risk mitigation 47, 48, 60–62, 78, 79, 84, 139, 182
  risk perception 30, 39, 52–55, 68–70, 76–78, 152, 157
  risk score 82, 83, 86, 87
  risk tolerance 39, 52–56, 72, 75–78, 151, 152, 157
  social risks 65
Rossy Yves 13

S

Sacred 60-minute 152, 157
  30-minute bubble 152, 155
safety briefing 15, 57, 60, 67–69, 72, 73, 86, 119, 126, 156, 157, 195
safety management system (SMS) 88, 98, 105, 111, 138, 139
  airshow-safety management system (ASMS) 45, 88, 89
safety observer 55, 73, 74, 80, 157, 175, 176
Scholes International Airport 187
science, technology, engineering, and mathematics (STEM) 208, 216
Serbinenko Anna 25
severity 46, 70, 71, 81, 83, 85–87, 127
Shoreham air show accident 1, 16, 31, 50, 90, 105, 118, 132, 136–138, 140
SHOWCENTER© 80–82
single-pilot resource management (SRM) 139
situational awareness 31, 49, 52, 64, 75, 82, 86, 109, 133–136, 137–140, 173, 175, 185, 188–190, 197, 213, 218, 224
Sknyliv air show accident 1, 38, 50, 90,
sleep 110, 156
  sleep deprivation 62
Spitfire aircraft 187–191
standard operating procedures (SOP) 2, 3, 48, 54, 60, 126, 174, 185, 190
Stearman aircraft 195, 196
stress 30, 31, 45, 51, 64, 72, 118, 123, 125, 133–136, 139, 145–150, 159, 161, 172, 188, 193, 196
  chronic stress 50, 145, 146
  high-stress environment 133, 173, 188
  high-stress situations 133, 139, 140, 149, 150, 180, 181, 183, 193, 197
  stress management 34, 136, 162
sustainability 27, 34, 206–208, 216, 217, 220–222
  environmental sustainability 27, 207, 217, 221, 222
  financial sustainability 205, 206
  sustainable aviation fuels (SAFs) 220

T

task 6, 30, 31, 51, 64, 82, 122–124, 126, 131, 135, 137, 144, 146, 147, 155, 161, 167, 170, 171, 173, 174, 176, 180, 188–190, 197

multitasking 170, 171, 173, 174
task saturation 188
task switching 170, 173, 174
teamwork 30, 31, 34, 66, 96, 100, 101, 130, 150, 178, 179, 195–198, 219
threat error management (TEM) 139
Tiano Frank 11
time compression 192, 193
Toponar Volodymyr 38
training 2, 5, 13, 14, 23, 31, 33, 37, 39, 48, 50, 54–56, 60, 62, 64, 65, 68, 69, 75, 76, 78–80, 83, 85–88, 100, 103, 107, 110, 112–114, 126, 130, 132–136, 137–140, 146–152, 160, 172–175, 177, 178, 180–182, 184–187, 193, 197, 213, 217, 219
aerobatic training 51, 217
flight training 60, 174
mindfulness training 144, 145, 150–152
pilot training 181, 194
risk perception training 68, 69
simulator training 174, 175
tunnel vision 193, 196
Turner Roscoe 37

U

UK Civil Aviation Authority 15, 16, 31
unexpected hazards 59, 60
United States Army Golden Knights 9
United States Hang Gliding and Paragliding Association (USHPA) 10
United States Navy Leap Frogs 9
University of North Dakota (UND) 216
unmanned aerial vehicles (UAV) 11, 16, 210, 212, 213, 225
upset prevention recovery training (UPRT) 60
urban air mobility (UAM) 210, 213, 214, 222

V

vigilance 3, 57, 64, 131, 188–190
virtual reality (VR) 31, 103, 139, 159, 210, 219, 222
visual limitations 136, 137, 139
visualization 34, 131, 146, 148, 151, 152, 158, 159, 161

W

Wagstaff Patty xii, 4, 18, 78, 79
walk the talk 146, 158–160
Warbirds 7, 8, 28, 64, 157, 187
wildlife management 194, 195
wing walker 120, 195–198
Women In Aviation International (WAI) 18, 24
workload 107, 120, 125, 135–137, 139, 188–190
Wright brothers 35, 36

Y

Yegorov Yuriy 38

Z

Zelazny Aerobatics Team 50
zero tolerance 55, 57, 58, 61, 108
Zeus F-16 Demo Team, Hellenic Air Force 6, 167, 168, 174

5

5B
5B domino effect 123
5B model 120, 121, 136–138
5M model 55, 57, 61–63

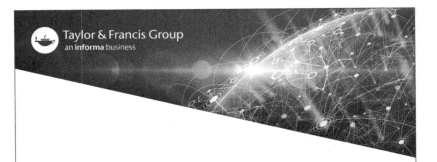

Printed in the United States
by Baker & Taylor Publisher Services